Opposing Viewpoints®

Energy Alternatives

Other Books of Related Interest

Opposing Viewpoints®

Energy Alternatives

Helen Cothran, *Book Editor*

Daniel Leone, *President*

Bonnie Szumski, *Publisher*

Scott Barbour, *Managing Editor*

OPPOSING
VIEWPOINTS®
SERIES

GREENHAVEN PRESS
SAN DIEGO, CALIFORNIA

GALE GROUP
™
THOMSON LEARNING

Detroit • New York • San Diego • San Francisco
Boston • New Haven, Conn. • Waterville, Maine
London • Munich

#47643323

Cover photo: © Photodisc

Library of Congress Cataloging-in-Publication Data

Energy alternatives : opposing viewpoints / Helen Cothran, book
    editor.
        p. cm. — (Opposing viewpoints series)
    Includes bibliographical references and index.
    ISBN 0-7377-0904-9 (pbk. : alk. paper) —
ISBN 0-7377-0905-7 (lib. bdg. : alk. paper)
    1. Power resources. 2. Renewable energy sources. I. Cothran,
Helen. II. Opposing viewpoints series (Unnumbered)

TJ163.2 .E4555 2002
333.79—dc21
                                                                2001040604
                                                                        CIP

Copyright © 2002 by Greenhaven Press,
an imprint of The Gale Group
10911 Technology Place, San Diego, CA 92127

"Congress shall make no law...abridging the freedom of speech, or of the press."

First Amendment to the U.S. Constitution

The basic foundation of our democracy is the First Amendment guarantee of freedom of expression. The Opposing Viewpoints Series is dedicated to the concept of this basic freedom and the idea that it is more important to practice it than to enshrine it.

Contents

Why Consider Opposing Viewpoints?

"The only way in which a human being can make some approach to knowing the whole of a subject is by hearing what can be said about it by persons of every variety of opinion and studying all modes in which it can be looked at by every character of mind. No wise man ever acquired his wisdom in any mode but this."

John Stuart Mill

In our media-intensive culture it is not difficult to find differing opinions. Thousands of newspapers and magazines and dozens of radio and television talk shows resound with differing points of view. The difficulty lies in deciding which opinion to agree with and which "experts" seem the most credible. The more inundated we become with differing opinions and claims, the more essential it is to hone critical reading and thinking skills to evaluate these ideas. Opposing Viewpoints books address this problem directly by presenting stimulating debates that can be used to enhance and teach these skills. The varied opinions contained in each book examine many different aspects of a single issue. While examining these conveniently edited opposing views, readers can develop critical thinking skills such as the ability to compare and contrast authors' credibility, facts, argumentation styles, use of persuasive techniques, and other stylistic tools. In short, the Opposing Viewpoints Series is an ideal way to attain the higher-level thinking and reading skills so essential in a culture of diverse and contradictory opinions.

In addition to providing a tool for critical thinking, Opposing Viewpoints books challenge readers to question their own strongly held opinions and assumptions. Most people form their opinions on the basis of upbringing, peer pressure, and personal, cultural, or professional bias. By reading carefully balanced opposing views, readers must directly confront new ideas as well as the opinions of those with whom they disagree. This is not to simplistically argue that everyone who reads opposing views will—or should—change his or her opinion. Instead, the series enhances readers' understanding of their own views by encouraging confrontation with opposing ideas. Careful examination of others' views can lead to the readers' understanding of the logical inconsistencies in their own opinions, perspective on why they hold an opinion, and the consideration of the possibility that their opinion requires further evaluation.

Evaluating Other Opinions

To ensure that this type of examination occurs, Opposing Viewpoints books present all types of opinions. Prominent spokespeople on different sides of each issue as well as well-known professionals from many disciplines challenge the reader. An additional goal of the series is to provide a forum for other, less known, or even unpopular viewpoints. The opinion of an ordinary person who has had to make the decision to cut off life support from a terminally ill relative, for example, may be just as valuable and provide just as much insight as a medical ethicist's professional opinion. The editors have two additional purposes in including these less known views. One, the editors encourage readers to respect others' opinions—even when not enhanced by professional credibility. It is only by reading or listening to and objectively evaluating others' ideas that one can determine whether they are worthy of consideration. Two, the inclusion of such viewpoints encourages the important critical thinking skill of objectively evaluating an author's credentials and bias. This evaluation will illuminate an author's reasons for taking a particular stance on an issue and will aid in readers' evaluation of the author's ideas.

It is our hope that these books will give readers a deeper understanding of the issues debated and an appreciation of the complexity of even seemingly simple issues when good and honest people disagree. This awareness is particularly important in a democratic society such as ours in which people enter into public debate to determine the common good. Those with whom one disagrees should not be regarded as enemies but rather as people whose views deserve careful examination and may shed light on one's own.

Thomas Jefferson once said that "difference of opinion leads to inquiry, and inquiry to truth." Jefferson, a broadly educated man, argued that "if a nation expects to be ignorant and free . . . it expects what never was and never will be." As individuals and as a nation, it is imperative that we consider the opinions of others and examine them with skill and discernment. The Opposing Viewpoints Series is intended to help readers achieve this goal.

David L. Bender and Bruno Leone,
Founders

Greenhaven Press anthologies primarily consist of previously published material taken from a variety of sources, including periodicals, books, scholarly journals, newspapers, government documents, and position papers from private and public organizations. These original sources are often edited for length and to ensure their accessibility for a young adult audience. The anthology editors also change the original titles of these works in order to clearly present the main thesis of each viewpoint and to explicitly indicate the opinion presented in the viewpoint. These alterations are made in consideration of both the reading and comprehension levels of a young adult audience. Every effort is made to ensure that Greenhaven Press accurately reflects the original intent of the authors included in this anthology.

Introduction

"Deregulation and competition should set the stage for an unprecedented national referendum on the future of America's energy policy—a vote up or down on nonpolluting 'renewable' energy sources: solar, wind, geothermal, biomass and hydroelectric power."
—Bill Paul, founder and publisher of Earth Preservers, an environmental newspaper for children

Computer screens go blank. Traffic lights go out. Air conditioners switch off, store lights darken, and factory machines grind to a halt. California's rolling blackouts—ordered in 2000 by the state's power managers in an effort to conserve dwindling electricity supplies—cost the state millions of dollars. The blackouts also reinvigorated the debate about the promise of alternative energy sources, such as wind and solar power, to help California and the nation solve a deepening energy crisis.

Many analysts blame California's energy crisis on the deregulation scheme that Californians voted for in 1996. However, most commentators agree that the state's energy problems can be traced to the environmental movement that began in the 1970s. At that time, environmental activists began pushing for increased regulation of industries—including power plants—that burned fossil fuels, claiming that such facilities polluted the air. While California's tougher standards did result in cleaner air, they also made the construction of power plants in the state much more expensive. As a result, many utilities stopped building new plants and instead began to buy power from out-of-state suppliers that operated without such strict environmental regulations. At that time, the prices that suppliers could charge utilities and the rates that utilities could charge consumers were regulated by the state, so the electricity market in California remained relatively stable, despite the dearth of new power plants.

In 1996, however, free-market advocates convinced Californians to vote for deregulation of the energy market. Since deregulation would end the monopoly that utilities had enjoyed, these advocates argued, electricity providers would have to compete with one another for business, which would lower prices and open up the market to alternative energy producers. These same advocates contended that consumers could buy "green power," that is, energy produced from renewable sources, thereby helping the environment. In 1998, deregulation in California went into effect.

Unfortunately, the deregulation scheme did not work as promised. Instead of going down, prices for electricity went up. For example, in 1999, the price per megawatt hour of electricity was thirty dollars. By 2000, the price had gone up in some areas to two hundred and fifty dollars per megawatt hour. The average consumer's electric bill more than doubled. However, supporters of deregulation contend that California's experiment failed not because deregulation was a bad idea but because the state did not deregulate completely. For one thing, the government mandated a surcharge on consumers' electric bills in order to help relieve the states' largest utilities of debts they had incurred years earlier for building expensive nuclear power plants. Paying off these "stranded costs" was intended to allow the utilities to compete with newer electric companies that had no such debts to pay off. Under this partial deregulation scheme, many consumers were forced to stay with their current energy supplier until they paid off that supplier's stranded costs. Only then could consumers switch to cheaper alternative suppliers. Supporters of deregulation contend that this law undermined competition and forced consumers to pay higher electric rates.

Not only did the price for electricity rise after deregulation, but the state experienced an energy shortage as well. Deregulation advocates point out that while the government had stopped controlling the price that out-of-state suppliers could charge California's utilities for power, it left the price cap that utilities could charge consumers intact. As a result, utilities paid more for electricity than they could sell it for. Soon utilities were unable to pay suppliers for power, and many suppliers stopped selling to California entirely. In 2001, California was forced to approve a 46 percent rate hike for consumers in an effort to save the utilities from bankruptcy. Eventually the government took over the purchase of electricity from out-of-state suppliers.

The promise that deregulation would result in a transition to alternative energy sources went largely unfulfilled as well. While many consumers liked the idea of buying "green" power, in fact, few actually did so. For one thing, many consumers were so confused by the plethora of energy suppliers, they simply stayed with their local utility. In addition, consumers discovered that electricity produced from wind farms and solar energy facilities—because their output of energy was small compared with facilities that burned fossil fuels—was more expensive than power provided by the big utilities.

Advocates of green power remain divided over deregulation's long term effect on the future of alternative energy sources. As the *CQ Researcher* explains, "the process of being able to choose one's power supplier is creating a burst of environmental awareness in some states that have adopted electricity deregulation plans." Conversely, *Issues and Controversies on File* points out that "competitive pressures may . . . force many utilities to abandon their use of renewable energy." Whereas in a regulated marketplace utilities were required to invest in alternative energy sources and were given subsidies to do so, no such rules or incentives typically exist in a deregulated market. The only incentive for utilities to invest in alternatives would be if customers proved willing to pay more for electricity from renewable sources, but so far, that has not been the case.

Many arguments about how to solve California's energy crisis center around the advantages and disadvantages of deregulation. However, other commentators—including president George W. Bush—point to California's problems as evidence that more power plants need to be constructed and currently protected federal lands need to be opened to oil, coal, and gas exploration. Other analysts argue that Californians need to conserve energy by shutting off lights, using energy-saving appliances, and installing more home insulation. Finally, many environmentalists insist that only a switch to alternative energy sources will provide a long-term solution to the power shortage that California faces. They claim that such energy sources are limitless and do not harm the environment the way that energy produced from fossil fuels does.

The solutions that California adopts to solve its energy crisis have profound implications for the nation. Indeed, over half of the states have deregulated their electricity markets, and some are experiencing problems similar to those in California. Furthermore, other regions—notably the Pacific Northwest—are also experiencing power shortages and rising rates, partly as a result of selling power to California, and partly due to droughts that reduce hydroelectric output. The authors in *Opposing Viewpoints: Energy Alternatives* debate whether or not alternative energy sources are the best solution to the nation's energy problems in the following chapters: Are Alternative Energy Sources Necessary? Is Nuclear Power a Viable Energy Alternative? What Alternative Energy Sources Should Be Pursued? and, Should Alternatives to Gasoline-Powered Vehicles Be Pursued? California's electricity crisis underscores the pressing need for long-term solutions in meeting the nation's growing appetite for energy.

CHAPTER 1

Are Alternative Energy Sources Necessary?

Chapter Preface

The United States depends on foreign producers for more than half of the oil it consumes. Much of the world's oil is produced by OPEC—the Organization of Petroleum Exporting Countries—which has the ability to manipulate supplies and prices worldwide. Former U.S. president Bill Clinton argued, "The nation's growing reliance on imports of crude oil and refined petroleum products threaten (*sic*) the nation's security because they increase U.S. vulnerability to oil supply interruptions." Many Americans worry that energy shortages could seriously disrupt America's new high-tech economy. Moreover, concerns about energy dependency have only deepened in recent years—U.S. reliance on foreign oil has increased from 35 percent in the 1970s to 56 percent in 2000.

One solution to America's energy dependence is to increase domestic production of oil. However, as critics of government restrictions on land use point out, regulations have closed many federal lands to oil exploration. Proponents of more domestic oil production want to see those restrictions lifted. Columnist Joseph Perkins contends that "to overcome the cartel's [OPEC's] market domination—which enables it to artificially inflate the price of crude any time it wants, driving up energy prices here in the United States—this country must markedly increase its domestic production of fossil fuels."

However, many commentators contend that more domestic oil production will seriously harm the environment by allowing drilling in ecologically sensitive areas. In addition, they point out that the nation's oil reserves are finite. These analysts argue that alternative energy sources can provide a better solution to America's dependence on foreign oil. They maintain that renewable energy sources, such as solar and wind power, are limitless and do not cause environmental degradation. In addition, many analysts argue that alternative energy sources have the potential to generate enough electricity to meet America's energy needs. Denise A. Bode, vice chairman of the Oklahoma Corporation Commission, argues, "As we put together a plan of action [to address U.S. dependence on foreign oil], conservation and alternative fuels must be part of our portfolio."

U.S. reliance on OPEC countries for much of the nation's energy certainly raises concerns about national security and the future health of the economy. Developing alternative energy sources is one solution proposed to help cut dependence on foreign oil producers. The authors in the following chapter debate whether or not alternative energy sources are necessary. To be sure, fears about energy scarcity will only increase in the future as the world's population expands and worldwide energy needs increase.

> *"An energy transition [to renewable energy
> sources] is . . . ecologically necessary, but it
> is also economically logical."*

A Transition to Renewable Energy Sources Is Necessary

Christopher Flavin

In the following viewpoint, Christopher Flavin argues that the United States must make a transition from burning environmentally damaging fossil fuels to the use of clean and renewable energy sources. He maintains that although renewable energy sources such as wind and solar power corner a minor share of the energy market today, they will become economically viable in the near future. Christopher Flavin is senior vice president at Worldwatch Institute, an organization that tracks the earth's well-being, and coauthor with Nicholas Lenssen of *Power Surge: Guide to the Coming Energy Revolution.*

As you read, consider the following questions:
1. According to Flavin, what is the legacy left by the last century's energy system?
2. How much of the world's energy do wind and solar power currently produce, according to the author?
3. In the author's opinion, what factors will influence the rate of the energy transformation from fossil fuels to renewable energy sources?

The age of oil has so dominated social and economic trends for the last 100 years that most of us have a hard time imagining a world without it. Oil is cheap, abundant, and convenient—easy to carry halfway around the world in a supertanker or across town in the tank of a family sport utility vehicle. From Joe Sixpack to the PhD energy economists employed by governments and corporations, we tend to assume that we will burn fossil fuels until they're gone, and that the eventual transition will be painful and expensive.

The Dark Legacy of Fossil Fuels

But if you turn the problem around, our current energy situation looks rather different: from an ecological perspective, continuing to depend on fossil fuels for even another 50 years—let alone the century or two it might take to use them up—is preposterous. As the new century begins, the world's 6 billion people already live with the dark legacy of the heavily polluting energy system that powered the last century. It is a legacy that includes impoverished lakes and estuaries, degraded forests, and millions of damaged human lungs.

Fossil-fuel combustion is at the same time adding billions of tons of carbon dioxide to the atmosphere each year, an inexorable escalation that must end soon if we are not to disrupt virtually every ecosystem and economy on the planet.

An energy transition in the new century is therefore ecologically necessary, but it is also economically logical. The same technological revolution that has created the Internet and so many other 21st century wonders can be used to efficiently harness and store the world's vast supplies of wind, biomass, and other forms of solar energy—which is 6,000 times as abundant on an annual basis as the fuels we now use. A series of revolutionary technologies, including solar cells, wind turbines, and fuel cells can turn the enormously abundant but diffuse flows of renewable energy into concentrated electricity and hydrogen that can be used to power factories, homes, automobiles, and aircraft.

A Growing Niche

These new energy conversion devices occupy about the same position in the economy today that the internal combustion engine and electromagnetic generator held in the 1890s. The key enabling technologies have already been developed and commercialized, but they only occupy small niche markets—and their potential future importance is not yet widely appreciated. As with the automobile and incandescent lightbulb before them, the solar cell and hydrogen-electric car are steadily gaining market share—and may soon be ready to contribute to a third energy revolution. They could foster a new generation of mass-produced machines that efficiently and cleanly provide energy needed to take a hot shower, sip a cold beer, or surf the Internet.

Ancient Energy Sources

Most of today's renewable and sustainable technologies are refinements of those used in one form or another for hundreds or thousands of years. We see pervasive evidence in the 2,000-year-old Roman baths, heated by the sun; in Persian windmills dating to 250 B.C.; in the thousand-year-old Anasazi cliff dwellings of the American Southwest, with their passive solar design; and in the use of geothermal pools for heating or medicinal purposes. What is relatively new in the last century is the use of renewable energy technologies to generate electricity. A prime example is the development at Bell Laboratories of modern solar electric (photovoltaic) cells, which were announced in 1954 and quickly became the primary power source for space satellites.

Mark C. Fitzgerald, *World & I*, March 1999.

Thanks to a potent combination of advancing technology and government incentives, motivated in large measure by environmental concerns, the once glacial energy markets are now shifting. During the 1990s, wind power has grown at a rate of 26 percent per year, while solar energy has grown at 17 percent per year. During the same period, the world's dominant energy source—oil—has grown at just 1.4 percent per year.

Wind and solar energy currently produce less than 1 percent of the world's energy, but as the computer industry long ago discovered, double-digit growth rates can rapidly turn a tiny sector into a giant. In the past two years, perhaps a dozen major companies have joined Royal Dutch Shell in announcing major new investments in giant wind farms, solar manufacturing plants, and fuel cell development. The "alternative" energy industry is beginning to take on the same kind of buzz that surrounded [Standard Oil president] John D. Rockefeller's feverish expansion of the oil industry in the 1880s—or [Microsoft president] Bill Gates's early moves in the software business in the 1980s. This January, stocks of solar and fuel cell companies suddenly jumped several-fold in a month, following the pattern of Internet stocks.

Profound Changes

The 21st century may be as profoundly reshaped by the move away from fossil fuels as the 20th century was shaped by them. Energy markets, for example, could shift abruptly, drying up sales of conventional power plants and cars in a matter of years, and influencing the share prices of scores of companies. The economic health—and political power—of whole nations could be boosted, or in the case of the Middle East, sharply diminished. And our economies and lifestyles are likely to become more decentralized with the advent of new energy sources that provide their own transportation network—for example, the sunshine that already falls on our rooftops.

How quickly the world's energy economy is transformed will depend in part on whether fossil-fuel prices remain low and whether the opposition of many oil and electric power companies to a new system can be overcome. The pace of change will be heavily influenced by the pace of international negotiations on climate change and of the national implementation plans that follow. In the 1980s, California provided tax incentives and access to the power grid for new energy sources, which enabled the state to dominate renewable-energy markets worldwide. Similar incentives and access have spurred rapid market growth in several European countries in the 1990s. Such measures have begun to overcome the momentum of a century's investment in fossil fuels.

Earth Day 2000—with its central theme, "Clean Energy Now!"—provides a timely opportunity for citizens to express their desire for a new energy system, and to insist that their elected officials implement the needed policy changes. If they do so, smokestacks and cars may soon look as antiquated as manual typewriters and horse drawn carriages do.

"Could it possibly be that the costs of generating electricity [using renewable energy sources] are higher?"

Renewable Alternatives to Fossil Fuels Are Unnecessary

Thomas Sowell

Thomas Sowell is an economist and a nationally syndicated columnist. In the following viewpoint, Sowell contends that energy shortages cannot be prevented by transitioning to renewable energy alternatives. For one thing, such alternatives to fossil fuels are expensive and too inefficient to power large states such as California, he contends. In addition, Sowell argues that governmental regulations cause energy shortages by making the construction of new power plants too expensive, and he claims that only by privatizing the energy industry can energy crises be averted.

As you read, consider the following questions:
1. How do price controls affect energy shortages, according to Sowell?
2. According to the author, what happens when governments regulate industries such as the electricity market?
3. In Sowell's opinion, why don't electricity industry workers support the idea of using renewable energy alternatives?

A s an economist, whenever I hear the word "shortage" I wait for the other shoe to drop. That other shoe is usually "price control." So it was no great surprise to discover, after the electric power shortage in California made headlines, that there were price controls holding down the price of electricity to the consumers.

Price Controls and Shortages

In the absence of price controls, a shortage is usually a passing thing. When prices are free to rise, that causes consumers to buy less and producers to produce more, eliminating the shortage. But when the price is artificially prevented from rising, the shortage is prevented from ending.

The electric power shortage in California is not unique. What is a new twist, however, is that there are no limits on how much the wholesale electric power suppliers can charge the utility companies that directly supply the consumer.

Since the utility companies have been paying more for electricity than they were allowed to charge their customers, they were operating in the red, and the financial markets are downgrading their bonds. Buying high and selling low is the royal road to bankruptcy, and bonds in a bankrupt company are not usually worth much.

There's No Free Lunch, No Free Electricity

Nor is it any great surprise that "consumer advocates" are denouncing the utilities for seeking a rate increase—or that politicians are proposing a small increase, completely inadequate to cover the cost of the electricity bought by the utilities. In the never-never land of California ideology, it is considered terrible if the public should have to pay the full cost of what it wants.

In California, prices higher than you like are attributed to "greed" or "gouging" and the answer is either more government regulation or having the government take over the utility company completely and run it. There are people who are old enough to know better who get their 15 minutes of fame by going on television and repeating the sophomoric slogans of their youth, back in the [hippie] days of Berkeley in the 1960s. And there are media people who take them seriously—or at least pretend to.

But just as there is no free lunch, there is no free electricity. And the idea that the government can run businesses at lower costs flies in the face of worldwide evidence that whatever enterprise politicians and bureaucrats run has higher costs. That is why even left-wing governments have been privatizing in recent years, even if this fact has not yet gotten through to those Californians who are still living with the ideological visions of their Berkeley youth.

Far from lowering the cost of producing electricity, government at all levels has for many years and in many ways been needlessly increasing that cost.

The Drawbacks of Renewables

Renewables are not without their drawbacks. Solar and wind farms cannot generate much electricity on cloudy or still days. As intermittent energy sources, they require vast systems to store the energy they produce, or must rely on the rest of the electrical system for backup. And despite federal subsidies to spur technological innovation, renewable sources have not become economical enough to seriously challenge fossil fuels in an open market.

As a result, renewables still accounted for just 8 percent of total U.S. energy production in 1995.

CQ Researcher, November 7, 1997.

Nothing forces prices up like restricting the supply. It has been years since anyone has built more electricity-generating facilities in California because the environmentalists, the courts, the state and local governments and assorted wackos have made it virtually impossible to build a hydroelectric dam, a nuclear power plant or a facility that uses coal or oil to generate electricity.

Renewable Energy Sources Are Too Expensive

There are all sorts of bright ideas for generating electricity by using sunlight or windmills. It never seems to occur to those who espouse these ideas to ask why people who have spent a lifetime working in the electricity industry do not share their enthusiasm for these schemes. Could it possibly be that the costs of generating electricity this way are higher?

There are already vast arrays of aging windmills in the hills leading out to California's central valley as monuments to the utopianism that seems to flourish in the Golden State. All that is needed is Don Quixote [writer Miguel de Cervantes' knight who pursued unreachable ideals].

Politics is supposed to be the art of the possible but, in California especially, it is often the art of the impossible. Somehow politicians must make it seem possible to get benefits without paying costs. But if we are too squeamish to build a dam and inconvenience some fish or reptiles, too aesthetically delicate to permit drilling for oil out in the boondocks and too paranoid to allow nuclear power plants to be built, then we should not be surprised if there is not enough electricity to supply our homes and support a growing economy.

The easy answer that is preferred is to use electricity generated outside of California—somewhere out in the real world beyond our borders.

| "The world is using up its geological
endowment of oil at a prodigious rate."

Accessible Oil Reserves Are Running Out

C.J. Campbell

C.J. Campbell argues in the following viewpoint that reserves of easily extracted oil are declining. He estimates that by 2003, half of the world's finite supply of conventional oil will have been extracted, and claims that discoveries of new deposits have been waning since the 1960s. Unfortunately, Campbell maintains that the transition to alternative energy sources and to unconventional oil reserves—which are difficult and costly to extract—will engender grave economic and political consequences. C.J. Campbell has spent a career in the oil exploration business and has written the book, *The Coming of the Oil Crisis*.

As you read, consider the following questions:
1. According to Campbell, how many barrels of oil are found for every barrel used?
2. Where does oil come from, according to the author?
3. In the author's opinion, for how long have human beings lived sustainably on Earth?

Excerpted from "Running Out of Gas: This Time the Wolf *Is* Coming," by C.J. Campbell, *National Interest*, Spring 1998. Copyright © 1998 by *National Interest*. Reprinted with permission.

I n the not-yet-named era in which we now live beyond the Cold War, a philosophically edged disagreement has arisen among our literati and social seers as to just how dangerous the world really is. Much of the discussion seems to turn on temperament or historic intuition rather than evidence. But of all the dangers cited by the more pessimistic among us—unconventional weapons proliferation, clashes of civilizations, global warming and environmental despoliation, rampant ethno-nationalism, and the rest—one almost never hears of the possibility of a major, world-wrenching energy crisis. And yet such a difficulty is at least as likely as any of these other would-be terrors. Given the false alarms that have been raised in the past, this assertion is certain to be received with considerable skepticism. It is an assertion I intend to justify in what follows.

Using Up the Earth's Endowment of Oil

The world is using up its geological endowment of oil at a prodigious rate, and that rate will increase as newly wealthy countries, particularly in Asia, enter the industrial phase of economic growth—as indeed they will, recent hiccoughs [in 1998] notwithstanding. At the same time, and despite astounding advances in the science of geology and in techniques of finding fossil fuel deposits, discovery rates of new oil reserves are falling sharply. For every four barrels used, only one is found. The lines of discovery, consumption, and extraction are bearing down on one another and will inevitably cross, probably in the year 2003; at that point, the world will pass its peak production of oil, meaning that more than half of the world's finite supply of conventional oil will have been extracted and consumed. When that happens—and 2003 is only five years away—both the developed and developing countries of the world will increasingly be faced with politically difficult and, ultimately, hugely expensive problems of adjustment. The international political problems facing the United States will be shaped in particular by the fact that as the world proceeds down the slope of oil production, five Middle Eastern countries will increasingly gain relative share. The last time that happened, in the 1970s when their share was about 36 percent, there was, in point of fact, hell to pay. Soon—in all probability by the year 2000—that share will stand at around 30 percent. . . .

Not Production, but Extraction

For most of the 20th century, the world has relied primarily upon an abundant supply of cheap oil to drive its economy. It is difficult to think of any aspect of modern life, from transport to agriculture, that does not depend upon it. Its depletion is therefore an immensely important subject. The beginning of wisdom concerning this subject is that, despite standard usage, we do not *produce* oil, we *extract* it. This circumstance alone implies certain immutable facts about the situation at hand and, given the reluctance to face those facts, it is necessary to repeat them in brief.

Oil comes mainly from the decomposition of algae and is preserved at the bottom of stagnant lakes and marine troughs under certain conditions of pressure and heat. Gas comes from plant remains. As the organic material is buried, it is at a critical point converted by chemical reactions into oil and gas. It then migrates through the rocks to fill such geological reservoirs and traps as may be physically reached. The conditions for the formation and preservation of oil occurred only very rarely in time and place in the earth's geological history. Much oil that was once trapped has since leaked away. One of the factors responsible for the abundance of oil in the Middle East is the widespread occurrence of salt, an excellent seal, above the reservoirs.

Almost all the oil produced to date, as well as what waits to be produced over the next twenty years or so, can be described as *conventional*. Conventional oil is that which flows at high rates and in good quality, much of it from giant fields found long ago. Ninety percent of oil produced today comes from fields more than twenty years old, and seventy percent from fields more than thirty years old. This is the kind of oil we know most about, and it has been responsible for the supply to which we have grown accustomed.

There is also non-conventional oil. It comprises the heavy oil and tarsand deposits of western Canada, Venezuela, and Siberia; oil from enhanced recovery achieved by changing the characteristics of the oil in the reservoir through steam injection and other methods; oil in hostile environments; oil in small accumulations; and oil of poor quality. The essential characteristic of all these types of non-conventional oil is that they are difficult and slow to produce. They may also be expensive to transport or refine.

Starting at Zero and Ending at Zero

We need to know how much conventional oil there is left and when its production will peak. Oil being a finite resource, production starts at zero and ends at zero. It reaches peak somewhere in between, more or less coinciding with the midpoint of depletion, when half the total endowment has been extracted. To assess the total endowment, we need to know how much has been produced, how much remains to be produced from known fields, and how many new fields await discovery.

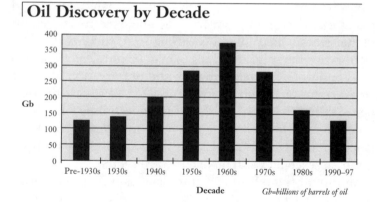

Oil Discovery by Decade

Decade — Gb=billions of barrels of oil

© C.J. Campbell. Used with permission of *National Interest*.

Oil & Gas Journal, the most widely used publication on the subject, reports that 807 billion barrels had been produced to the end of 1997, almost half of that since the shocks of the 1970s. Reported reserves in known fields are 1.020 trillion barrels, but, as shown below, there is good reason to reduce this estimate to about 830 billion barrels of median probability.[1] If these numbers are reasonably accurate, it

1. "Annual Production Report," *Oil & Gas Journal*, December 29, 1997.

means that approximately 1.635 trillion barrels have been discovered to date (produced + reserves). The graph on the previous page, which documents the number of gallons of oil discovered per decade, shows that peak discovery occurred during the 1960s. Extrapolating this historical discovery pattern suggests a total endowment of about 1.8 trillion barrels, which is in line with the trend of some twenty-five published estimates over as many years, as well as with other analytical techniques. That means, in turn, that about 165 billion barrels remain to be found, and about 995 billion barrels remain to be produced.

Decreasing Discovery Rates

This discovery plot makes a very compelling argument about the resource base of the earth's conventional oil. The discovery rate over the past thirty years has fallen, despite all the technological advances and the huge increase in knowledge gained from worldwide exploration. First came onshore exploration, which found the larger basins and most of the giant fields, which together hold some 70 percent of all known oil. Then offshore development was opened, effectively by the development of the semi-submersible rig in the 1960s. Discovery there, too, has now peaked. Attention is now turning to deep offshore deposits, but there are only a few deepwater areas that are geologically promising or indeed possible. To date, only about 25 billion barrels have been found in deep water—namely, about one year's world supply, and that from the more promising tracts. More can be expected, but not enough to defer peak by any significant period.

Some people have thought that great new possibilities for discovery would flow from the opening of the former Soviet Union after the fall of communism. It turns out, however, that Soviet explorers were as intelligent as their Western counterparts and were able to find the larger basins and most of the giant fields. Although they were working with rather primitive technology, it was not more primitive than that available to the West in those earlier years when most of its oil was found; after all, reserves in the Caspian area were discovered even before Colonel Drake drilled his famous well in Titusville, Pennsylvania in 1859.

In fact, too, the offshore extensions of this geologically productive area are well known and have been for many years. On the shores of the northern Caspian lies the giant but very difficult Tengiz Field, with reserves of about 8 billion barrels. There is some possibility that other similar finds may be made offshore, although there are doubts whether the critical salt seal is intact. There are also fears that the structures may contain gas and not oil. Additionally, there remain large, poorly known, and barely explored basins in the Siberian Arctic and offshore, but the extreme environment places such oil as they may contain in the non-conventional category. Apart from the resource constraints, doing business in the former Soviet Union has proved difficult for most foreign companies, so that even if its potential is greater than estimated, not much new oil is likely to find its way onto world markets for a long time to come. At best, Caspian oil will be of North Sea proportions—three years' world supply—and while this will help, it does not change the longer term picture very much.

The Battle over Oil

Very quickly the fight will be over the dwindling petro-reserves. . . . There will be winners, to be sure. Those nations rich in oil and strong in resolve will get their energy fix. America will thrive, so too can Britain and South Africa. Other European Union members will fare according to their ability to command and pay for it (and not in euros, thank you). Much of the social safety-net that defines the industrial West will be at risk.

There will be losers too. Some are already visible on the horizon. Russia, which has lost control of its own energy subsidies, is imploding. Japan, which must ship in each barrel of economic growth it utilises, is adrift. Even nations which have oil—one thinks of Indonesia and Nigeria—must show they can control it, lest it be poured down the drain of civil strife.

James Srodes, *Spectator*, August 29, 1998.

In short, the world has now produced almost half of its geological endowment of conventional oil. The earth has now been so extensively explored that virtually all the prolific producing trends have been identified. Geophysical techniques have also improved radically so that it is possible to map in very great detail the traps for oil that occur within the productive areas. There is, it so happens, a natural geological polarity about oil, reflecting the rare conditions of its formation: it is either there in abundance or it is not there at all. . . .

The End of Cheap Oil

Man arrived on the planet some two million years ago and lived sustainably on it until about 1850, when he numbered about one billion. Since then he has multiplied six-fold and the growth in oil extraction has increased in direct proportion to it. Since oil has provided abundant cheap energy to sustain the human population increase—particularly over the past half century—it can hardly be doubted that the imminent peak of oil production will mark the beginning of a historic discontinuity that will change the world we know and the energy economics that drive it. We are not running out of oil, just out of the cheap abundant supply on which we have come to depend.

There are many long-term solutions to the problem engendered by the depletion of cheap oil in terms of renewable energy sources and, above all, of using less of the endowment we have. Our main concern, however, must be about the transition from the era of cheap oil to whatever replaces it, a transition that will bear with it many grave economic and political consequences. These cannot be avoided entirely, but they can be mitigated with a modicum of planning. A proper understanding of the resource base is critical to any sensible political response. But a proper understanding seems to be precisely what governments are lacking, and these days are laboring actively to avoid having. Fundamentally the problem is not the greed of a few politically recalcitrant Middle Eastern countries that were especially favored during the Jurassic period—something that in the last resort could be solved by sending in the Marines. Rather, it is the immutable physics of the reservoir.

| *"The lesson of the last quarter century is that oil reserves are growing."*

Oil Reserves Are Not Running Out

Sarah A. Emerson

Sarah A. Emerson is director of Energy Security Analysis Inc., an energy research firm. In the following viewpoint, Emerson contends that the world's oil reserves are growing. She maintains that although oil deposits are continually being extracted from the Earth, technological innovation has enabled the oil industry to locate and extract from new reserves worldwide. She claims that governments are misinformed in creating energy policies based on the assumptions that oil reserves are diminishing and are located in the politically unstable Middle East.

As you read, consider the following questions:

1. What did Harold Hotelling argue about the future of the world's oil reserves?
2. How far does the United States fall behind the oil industry's recommended reserve-to-production ratio, according to Emerson?
3. According to the author, what factors contributed to the 27 percent increase in reserve estimates in 1987?

Excerpted from "Resource Plenty: Why Fears of an Oil Crisis Are Misinformed," by Sarah A. Emerson, *Harvard International Review*, Summer 1997. Copyright © 1997 by *Harvard International Review*. Reprinted with permission.

C oncerns over the "energy security" of an oil-consuming country arise from two commonly held assumptions. The first is the general assumption that global oil supplies are finite and dwindling. The second and more specific assumption is that most of the remaining oil is located in politically unstable regions such as the Middle East. Since the first oil crisis in 1973, these two assumptions have shaped energy policy in the oil-consuming countries of the Organization for Economic Cooperation and Development (OECD). The United States has essentially replaced fuel oil with natural gas in industrial consumption and electric power generation. In Europe, gasoline is so heavily taxed that the price at the pump is as much as five times higher than the price at the refinery gate. European countries have also compelled their refiners and importers to hold a minimum level of petroleum product inventory, while the United States, Japan, Germany, and, more recently, South Korea have built substantial government stockpiles. These and other policies have promoted conservation and now provide a buffer stock that will lessen the short-term price impact of a supply disruption.

Changing Assumptions

Though the issue of energy security remains a perennial concern of oil-importing countries, the global oil market is very different today than it was in the 1970s. Public policy must take into account the fact that the nature of energy security concerns has changed as the motivating assumptions that spawned those concerns have changed. Most importantly, the starting assumption of global resource scarcity is grossly exaggerated in public policy discourse: not only does output from mature producing regions continue to exceed expectations, but the exploration and production frontier continues to expand across countries as geographically and economically disparate as Vietnam, Turkmenistan, Chad, and the United States.

It would be foolhardy to contend that there is enough oil in the ground to meet global demand growth for the next 100 years, but energy policy is not formulated to address 100-year problems. Its horizon can be as near as the next election or as far as the foreseeable future, but even the latter is much closer to 20 years than 100. For any energy security policy to be effective over the next 20 years, it must recognize the resource plenty—as opposed to the resource scarcity—that results from the free flow of capital and technology across an increasingly global resource base. That does not mean that importing countries should ignore the question of energy security, but that policies should go beyond "getting out of oil" or "making the world safe for oil production." Importing countries should consider alternative approaches that encourage safe and efficient development of this abundant resource, especially from regions other than the Middle East. In the process, energy security policy can capitalize on resource plenty and lessen, if not postpone, overwhelming dependence on Middle Eastern oil.

The Theoretical Paradigm

The idea that the world is running out of oil was foreshadowed by Thomas Malthus' contention that the more the world consumes its resources, the less it has left. More recently, in energy circles, the issue has been encapsulated in the concept of "resource scarcity," a term closely associated with Harold Hotelling's 1931 paper, "The Economics of Exhaustible Resources," which has shaped perceptions of oil supply ever since. Hotelling's fundamental argument was that the future price of oil is an inclining curve, largely because the volume of oil in the ground is a finite and fixed stock. Therefore, as each barrel is produced and consumed, remaining barrels become dearer and more expensive.

In the intervening decades, however, Hotelling's premise has been modified if not refuted by subsequent analyses, most notably that of Morris Adelman of the Massachusetts Institute of Technology, who succinctly points out that instead of fixed stocks of resources, there are only flows of reserve-additions. In other words, oil reserves in the ground are not a stock but a flow. On one hand, oil is consumed, which diminishes reserves. On the other hand, the oil industry spends billions of dollars each year exploring for oil and, in the process, adding to reserves. Adelman goes on to say that the industry's objective is to maintain a volume of reserves that is equal to at least 15 years of production, a reserve-to-production ratio of 15. The vast majority of oil producing countries have reserve-to-production ratios far greater than 15, although some of the most mature producing regions, such as the United States and the United Kingdom, have ratios closer to 10.

Phases of Growth in Non-OPEC Supply

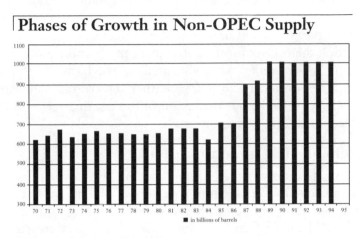

■ in billions of barrels

BP Statistical Review of World Energy

According to Adelman, replacement cost is the measure of scarcity. He posits that, "all else being equal, the replacement cost of any mineral should constantly increase over time, and the price with it. First, the average size of newfound deposits should constantly decrease. The biggest would be found first even by chance, let alone by design. Second, the better, (i.e., lower cost) mineral should be used up first." Adelman then responds to his own characterization: "Yet prices of minerals have not risen. Practically all prices have been flat or actually declining in the long run. . . . Mineral depletion is in fact an endless tug of war: diminishing returns versus increasing knowledge. So far, the human race has won big."

That victory is apparent in the historical growth of proven oil reserves. The accompanying graph of reserve estimates since 1970 shows that reserve additions more than offset production almost every year during this 25-year period. This fact alone suggests that oil is becoming more rather than less plentiful. Despite this result, we should put only so much faith in estimates of oil reserves, for the graph also illustrates the political manipulation of official reserve estimates. The hefty 27 percent increase in 1987 followed the 1986 oil price collapse and the end of Saudi Arabia's tenure as swing producer. Most of the increase came in the official reserve estimates for Venezuela, the United Arab Emirates, Iran, and Iraq. All four countries inflated their reserve estimates not only as part of the ensuing battle for market share, but also to position themselves for the subsequent production quota system in which quotas were allocated on the basis of reserve size. The lesson of 1986–1987 is that reserve estimates are not always reliable, but the lesson of the last quarter century is that oil reserves are growing.

Technological Innovation

Adelman's theory is very compelling. As long as the price of oil exceeds the cost of exploration, development, and extraction, companies will continue to invest in adding to their reserves. The main factors pushing costs down are increases in knowledge and improvements in technology. The resource scarcity argument, therefore, only works if additions to reserves dry up. This will only happen if the industry stops investing, which will only occur if replacement costs exceed price. To believe that replacement costs will exceed price, one must bet against technological innovation. It is the inexorable march of technological development that has allowed the industry to find and produce more and more oil in mature regions at lower and lower cost, while at the same time opening up remote and hostile oil-producing frontiers. The first step in recognizing resource plenty is to abandon Hotelling's fairly simplistic assumption of oil as a fixed stock in favor of Adelman's more subtle characterization of oil reserves as a flow, a flow dependent on continual adaptation and innovation to keep replacement (or production) costs below price. . . .

Resource Plenty as a Foundation for Policy

The concept of resource scarcity is not an appropriate underpinning for energy security policy. Resource plenty, which focuses the policymaker's attention on the future instead of the past, is a much more compelling basis on which to form policy. Given this reorientation in assumptions, policymakers can rank their security concerns and policy responses. If weathering a temporary supply disruption is the objective, then building and maintaining an emergency stockpile is the most direct solution. If the objective is to safeguard the economy against the implied threat of exhausting oil reserves and the resulting price rise, then proper recognition of the implications of resource plenty is essential. The solution is to fashion policy that promotes the efficient and safe development and use of this abundant resource. New technologies in the production and consumption of other energies may one day provide competitive alternatives to oil. Energy policy between now and then should do everything it can to unlock the earth's resource plenty. This is especially true in addressing the concomitant concern that ultimately the oil left will all be in the politically unstable Middle East. The more policy can unshackle the enormous resources in countries outside of the Middle East, the longer ultimate dependence on Middle Eastern oil can be postponed.

| *"Climate change is arguably the greatest threat facing humanity. Society's addiction to fossil fuels . . . is driving us relentlessly down a highway of self-destruction."*

The Fossil Fuels Industry Is Destroying the Environment

David Cromwell

David Cromwell argues in the following viewpoint that society's reliance on the combustion of fossil fuels for energy threatens to destroy the environment. He contends that oil companies—supported by government tax benefits and subsidies—have established a centralized power base from which they can dictate future energy policies that will benefit their industry at the expense of the environment. Cromwell recommends transitioning from a fossil fuels energy economy to a locally controlled, solar-based one, which will not harm the Earth. David Cromwell is an oceanographer, associate director of www.MediaLens.org and author of "Private Planet" (Jon Carpenter Publishing).

As you read, consider the following questions:
1. According to the Royal Commission on Environmental Pollution, by what percentage will carbon dioxide emissions have to fall within the next fifty years to positively affect the climate?
2. According to Cromwell, how much has the world's largest solar company spent on its solar energy division as compared to oil exploration?
3. What measures do "no regrets" policies include, according to the author?

From "The Climate Problem," by David Cromwell, *Z Magazine*, September 2000. Copyright © 2000 by the author. Reprinted with permission.

In their 1996 book *Who Owns the Sun?*, solar energy campaigners Daniel Berman and John O'Connor rightly declared that "democracy is a false promise if it does not include the power to steer the energy economy." It's a crucial point that not even Greenpeace and Friends of the Earth appear to have grasped; should we really be leaving it to the oil companies to create the solar revolution?

Big Oil

Climate change is arguably the greatest threat facing humanity. Society's addiction to fossil fuels—hard-wired by corporate greed and government handouts to the fossil fuel industry in the form of tax benefits and subsidies—is driving us relentlessly down a highway of self-destruction.

In the United Kingdom (UK), the Royal Commission on Environmental Pollution has just told the British government in a new report that carbon dioxide emissions must fall by 60 percent in the next 50 years if there is to be any realistic possibility of even "a tolerable effect on the climate." But how likely are such "huge cuts" while transnational corporations dictate how society produces and consumes energy? According to the San Francisco-based Transnational Resource and Action Center (TRAC), "Big Oil's long-term strategy is still dictated by the urge to explore." New exploration as well as oil or gas pipelines threatens the survival of people in the Amazon basin, Southeast Asia, and North America. BP Amoco, the world's largest solar company, is committed to spending $5 billion in the next 5 years on oil exploration and production in the sensitive environment of Alaska alone. This dwarfs the trifling sum of $45 million recently spent on its solar business division.

Meanwhile, Shell proudly proclaims that it is "focusing [its] energies on developing [renewable energy] solutions" even as its annual reports project fossil fuel growth and depict maps highlighting the global reach of its oil and gas enterprises. Shell's investment in renewables is only 10 percent of the oil giant's spending on hydrocarbon exploration ($1 billion annually), 0.8 percent of its global investment ($12 billion), and only 0.06 percent of its global sales ($171 billion): a drop in the barrel, in other words. Other companies such as the combined Exxon-Mobil, the world's largest oil corporation, are doing even less to develop renewables.

In the global economy, the unsustainable expansion of corporate activities into ever-larger markets means that there is an almost irresistible force driving the formation of mega-companies of all types. Growth demands further growth, and if companies do not expand in today's "internationally competitive" markets they stagnate and die. Smaller enterprises are swallowed up whole or trampled underfoot in the stampede to maintain or increase returns on short-term investment, or to repay loan capital. The business of generating energy is no different in this respect to other industrial operations; there is an inherent trend away from small-scale, community-based enterprises towards large-scale, centralized operations. It should therefore come as no surprise that oil companies are engaged in a frenzy of mergers. Describing the ongoing BP Amoco merger, The *Independent* newspaper in London coolly reported that, "the existing cost reduction plan involves 10,000 redundancies, of which 6,000 have already been achieved." At Exxon and Mobil, job losses will exceed 9,000. As TRAC notes, Exxon had already been cutting jobs at the rate of 4 percent every year for over a decade.

Transition to Renewable Energy Sources

Rather than pursuing such a destructive energy policy—in which corporations continue to overload the atmosphere with global-warming gases, destroy jobs, and damage sensitive ecosystems—society could be using local renewable energy sources. These come in many forms: wind, wave, solar, geothermal, small-scale hydro, and biomass fuels. Some of these are available at every location around the globe. Consequently, small-scale decentralized economies would be able to make use of a range of local energy sources for local needs. On the other hand, large industrialized economies with urbanized centers are locked into centralized power sources that convert fossil fuel or nuclear power into electricity, which is then transmitted over hundreds or even thousands of miles. This is extremely wasteful: two-thirds of the energy in fossil fuels is lost in the production and transmission process. Electricity is an indefensible luxury for 90 percent of our energy uses. Lighting and heating homes, for example, can be made much more energy-efficient by adopting "passive" solar building designs, low-energy lights, and tight insulation.

The United States Is a Big Polluter

The problems of fossil fuels—toxic spills, mining waste, acid rain, smog, etc.—haven't gone away. Meanwhile, a new problem has emerged: global climate change, with its multiple threats of rising sea levels, disrupting agriculture, increasing weather-related disasters and spreading infectious diseases. The scientific consensus is that climate change is happening, and its chief source is carbon dioxide released by the combustion of fossil fuels. The United States accounts for about a quarter of the world's energy consumption, so it's no surprise that this country also is responsible for 24 percent of carbon dioxide emissions—the largest source of which is power plants.

Eric Weltman, *In These Times*, February 7, 2000.

Energy efficiency is vastly underexploited. U.S. journalist Ross Gelbspan points out that "as a bridge to a new energy era," the economics panel of the UN Intergovernmental Panel on Climate Change has identified a number of steps, called "no regrets" policies. At virtually no cost, these could reduce greenhouse gas emissions by around 20 percent. They include such simple steps as implementing known efficiency and conservation techniques, planting more trees (to absorb carbon dioxide), and instituting international standards for energy-efficient appliances. Such measures should be encouraged at the same time as a switch to green energy. Removing fossil fuel and nuclear tax credits and subsidies, which currently promote the destruction of the global environment, and diverting them to windmill farms, home-based fuel cells, photovoltaic panels, and hydrogen fuel plants would provide the necessary boost to propel renewable energy into the big league of global industry. Renewable energy analyst Scott Sklar estimates that for every million dollars spent on oil and gas exploration, only 1.5 jobs are created; for every million on coal mining, 4.4 jobs. But for every million spent on making solar water heaters, 14 jobs are created. For manufacturing solar electricity panels, 17 jobs. For electricity from biomass and waste, 23 jobs.

A De-Centralized, Solar-Based Economy

In modern "civilization," the population tends to cluster in large cities in which a high-consumption lifestyle is encouraged. Profligate energy use, international trade, and the concentration of millions of people in urban centers are therefore intimately linked. This is why a decentralized, solar-based economy must go hand in hand with a revitalized locally based democracy; one cannot succeed without the other. What would such a society look like? Based on suggestions presented by Berman and O'Connor in *Who Owns the Sun?*, a blueprint for a solar society would incorporate:

- Public ownership of energy—just as is the case with water or schools in some countries and American states
- Massive investment in renewable energy technologies and building design, by diverting tax breaks and subsidies from fossil fuel and nuclear energy
- Access to loans, tax credits, and rebates for photovoltaics, solar water heating, wind, and small-scale hydro generators, and other forms of renewable energy-generating and energy-saving technologies
- Net metering (i.e., monitoring electricity flows) and rate-based incentives, so that independent home- and business-based electricity producers are paid the same price for electricity they supply to the grid as they would be required to pay for the grid power if they used it
- Partnerships between industry, governments, and local communities to oversee the new green industries, in order to ensure that the public knows what is being produced in a factory, by what means, and how any wastes and by-products will be managed
- New government legislation to ensure that all this is carried out

Local Control

None of the above will happen if we leave it to the giant oil corporations to tinker with solar renewables—as Shell and

BP Amoco are doing—while they bulldoze ahead with exploration and production of new oil and gas reservoirs. Citizen control over a decentralized solar economy is in direct competition with the profit imperative of such large companies. The present policy of governments and mainstream environmental organizations is to leave it up to fossil-fuel corporations and big utility companies to bring about a solar revolution. As Berman and O'Connor warn: this will "guarantee that the coming Solar Age will arrive a century behind its time, and that it will be every bit as autocratic as today's fossil-fuel economy." Decentralized renewable energy directed by local communities will only be won at the expense of the private energy monopolies that are currently engaged in cutthroat competition to protect and expand their share of the energy market. War-like metaphors abound in company rhetoric. Earlier this year, Shell group chair Mark Moody-Stuart glowingly described his company as "a great fleet of destroyers and torpedo boats." It's time to scuttle this fossil-fuel armada and launch a new fleet of solar-driven vessels fit for the 21st century.

"There is little reason to believe that industrial emissions of carbon dioxide and other greenhouse gases will produce a climatic catastrophe."

Environmentalists Have Exaggerated the Dangers of Burning Fossil Fuels

Jonathan H. Adler

Jonathan H. Adler is director of environmental studies at the Competitive Enterprise Institute, an organization that advocates the removal of environmental regulations, and the author of *Environmentalism at the Crossroads*. In the following viewpoint, Adler maintains that there is no scientific evidence that the combustion of fossil fuels leads to global warming. He warns that any treaty limiting the use of fossil fuels, if based on the false assumption that carbon dioxide emissions harm the environment, will raise energy prices, harm economies worldwide, and condemn developing nations to poverty.

As you read, consider the following questions:
1. According to Adler, what do satellite records and weather balloons indicate about the change in the Earth's temperature since 1979?
2. In the author's opinion, why did Hurricane Andrew result in heavier financial losses than the storm that struck Florida in 1926?
3. How many more highway fatalities will occur per year if the government mandates an increase in CAFE standards, according to the author?

Spectacular stories about melting polar ice caps, rising sea levels, severe storms and outbreaks of disease may make for banner headlines and increased book sales, but they are not the basis of sound policy. The more we learn about the Earth's climate and the human impact upon it the less we have to fear. There is little reason to believe that industrial emissions of carbon dioxide and other greenhouse gases will produce a climatic catastrophe.

False Apocalypse

Regrettably, many world leaders have taken these apocalyptic projections to heart and are preparing to sign an international treaty in December, 1997 in Kyoto, Japan, to reduce greenhouse-gas emissions.[1] The Clinton administration is on record calling for binding emission restrictions on developed nations, and there is increasing pressure for developing nations to accede to the same. Yet, while the risks of global warming are highly uncertain, the costs of global-warming policies are not. Signing a global-warming treaty would have dramatic consequences for the world's economy and the American way of life.

In mid-July, 1997, President Bill Clinton proclaimed that global warming no longer is a theory, but a fact. He should tell that to the planet's atmosphere. Precise measurements of the Earth's temperature obtained by satellite have detected no warming since the beginning of such measurements in 1979. In fact, the satellite record shows a slight, but statistically significant, cooling. Data from weather balloons confirm the satellite findings.

Average global temperatures have increased by approximately one-half of 1 degree Celsius since 1881. But this is not likely due to human activity, since most of the increase occurred prior to World War II, *before* the increase in industrial emissions of carbon dioxide.

If the Earth is not getting warmer, it is implausible to argue that various climatic changes are being caused by human activity. Computer models may predict significant climate

---

1. Former president Bill Clinton signed the Kyoto agreement in 1997, but the U.S. Senate refused the ratify the treaty.

changes in the future, but they have failed accurately to predict the present, as documented by recent reports in *Science*, the *Bulletin of the American Meteorological Society* and elsewhere. Moreover, peer-reviewed studies indicate that even were warming to occur, the parade of cataclysms trotted out by environmental activists are unlikely. For instance, against the claim of an increased threat of melting polar ice caps, a study just published in the prestigious journal *Nature* indicates that global warming would be just as likely to produce ice-cap thickening.

Outrageous Assertions

Ross Gelbspan argues that global warming will produce an increase in tropical storms. "Between 1990 and 1995, 16 floods, hurricanes and storms destroyed more than $130 billion in property—and caused deaths, homelessness and psychological damage for the many victims of those catastrophes," he writes in *The Heat Is On*. He blames these events—and the record-level insurance-company payouts that resulted—on the emission of carbon dioxide from the burning of fossil fuels.

Contrary to Gelbspan's outrageous assertions, there is no scientific basis upon which to blame recent hurricane activity or storm damage on human activity or global warming. Scientific reviews of storm data cannot find any correlation between warmer temperatures and increased hurricane activity. If anything, the existing data show a slight decline. William Gray of Colorado State University, one of the foremost hurricane experts in the world, insists that any increase in hurricane activity during the last few years is the result of natural variability, not human-induced climate change.

Increases in insurance-company losses have more to do with the unprecedented level of coastal development than the greenhouse effect. Hurricane Andrew was certainly a whopper. It caused an estimated $30 billion in damage in southern Florida. Yet Andrew hardly was the greatest storm ever to hit the Miami area. The storm that struck in 1926 was significantly worse. Research by David and Stanley Chagnon confirms that increases in storm-related property losses have more to do with increased development and the inflation of property values than significant variations in weather patterns.

It certainly is possible that the continued burning of fossil fuels eventually could produce global warming at some point in the future. Yet this fact in and of itself is no cause for alarm. The essential question is not whether the Earth's climate may change—nature is in constant flux—but what an increase in global temperatures would mean for life on the planet. The fact is, the Earth's climate has changed dramatically during the course of human history. But one should not assume that any climatic change is inherently bad.

Disastrous Global-Warming Policies

While global warming is highly uncertain, the effects of global-warming policies are not. Dramatic restrictions on energy use would have severe economic effects in this country and throughout the world. A review conducted at the Department of Energy's Argonne National Laboratory near Chicago determined that such a policy would increase energy prices dramatically, threatening the viability of energy-intensive industries, including aluminum, steel and paper. Is it any wonder that the AFL-CIO [labor union] is among the most vociferous critics of the rush to sign a global-warming treaty?

"Any plan that makes energy less abundant significantly will lower living standards," notes Frances Smith, executive director of Consumer Alert, a national consumer organization. Smith notes that consumer households account for about 25 percent of U.S. energy use. As a result, consumers will bear the brunt of a global-warming treaty. A draft study for the White House concluded that simply stabilizing emissions at 1990 levels by the year 2010—far less than Undersecretary of State Tim Wirth's stated goal of 70 percent reductions—would produce a significant increase in energy prices, equivalent to a 26-cent-per-gallon tax on gasoline and 2-cent-per-kilowatt-hour tax on electricity.

The costs of a treaty will hit more than consumers' pocketbooks, however. For instance, one policy favored by the Clinton-Gore administration to reduce greenhouse-gas emissions is an increase in Corporate Average Fuel Economy, or CAFE, standards for automobiles. This would be a disaster.

An increase in CAFE standards from their current level of 27.5 miles per gallon to 45 miles per gallon, as some have proposed, would force automakers to make smaller and lighter cars than they otherwise would produce. The result would be as many as 1,800 additional highway fatalities per year, on top of the 2,000 to 4,000 deaths that current CAFE standards already produce, according to a Harvard-Brookings study.

Fossil Fuels Save Lives

Plentiful energy, which is produced almost entirely from fossil fuels, has been responsible for a great improvement in the lives of men and women. It has reduced hardship, improved food production, and made transportation affordable for the common man. Everywhere on the globe, people are living longer, healthier lives than ever before.

Thomas Gale Moore, *Hoover Institute Newsletter*, Winter 1998.

Some argue that the United States and other nations can reduce emissions of carbon dioxide and other greenhouse gases at little cost to the economy. This simply is not the case. Reducing emissions inherently requires government taxes or regulations to control energy use, which will result in higher prices to consumers and produce economic dislocations. Whether the preferred control mechanism is some form of energy tax or emission quotas, the impact on consumers will be the same.

Alternative Energy Sources Are Expensive

Fossil-fuel energy is used so heavily because it is relatively cheap and abundant. There are no ready substitutes for most applications. Environmentalists have promised for decades that economically viable alternative-energy sources are "just around the corner." The world still is waiting. If such technologies were ready for mass production, there would be no need for government mandates or subsidies to put them in place, as they would be able to compete in the open market. Government research and technology-promotion programs sound good on paper, but they do not produce. The governmental takeover of entire industries to finance the development of alternative-energy sources, as proposed by Gelbspan in his book, is a recipe for economic disaster. Industrial policy has been a failure the world over, even in Japan. Painting it green will not make it work any better.

Even if there were a "free" way to reduce emissions as much as 30 percent, as Gelbspan has suggested, it would not matter. If the computer-model predictions of global warming are accepted, such reductions are too small and insignificant to have any effect on the climate. Averting the threat of global warming would require emission reductions two to three times that amount, and there is no economic study that suggests such reductions can be achieved at little to no cost.

Not only would the United States have to impose dramatic restrictions on energy use, these restrictions would have to be imposed throughout the world to be effective, as most of the growth in emissions in future years will come from developing nations such as Mexico, Brazil, India and China. Yet, as Gelbspan acknowledges in his book, "Those countries can barely afford to feed their people, let alone finance energy transitions." The greatest environmental threats faced by the peoples of the Third World are not caused by global warming, but by inadequate supplies of safe drinking water and sanitation. The World Health Organization estimates that unsafe water contributes to 3 million to 5 million deaths in developing countries each year. Imposing limitations on the use of energy in the poorer nations will condemn them to a future of poverty and despair.

Yet even if developing nations are excluded from a global-warming treaty, they still will feel its effect. Reducing economic growth in wealthier nations will slow economic growth the world over and cause export markets to contract. In addition, "the markets [in developed nations] to which these developing countries sell a large share of their exported goods will shrink, so most developing countries also would be harmed by the adoption of emission limits" on industrialized nations, according to economist David Montgomery of Charles River Associates. The impact of any treaty restricting greenhouse-gas emissions will be experienced worldwide.

The more we learn about the risks of global warming, the more it appears that they are outweighed by the risks of global-warming policy. Dramatic restrictions on the use of fossil-fuel energy will curtail economic growth in this country and devastate the developing economies of the Third World for no demonstrable environmental benefit. Lowering living standards worldwide with an uncertain solution to an uncertain problem is not merely imprudent, it is wrong.

"Ordinary people can help [stop global warming] immediately by becoming more energy efficient."

Conservation Can Help Solve Energy and Environmental Problems

Bob Herbert

Bob Herbert contends in the following viewpoint that energy conservation can decrease society's reliance on fossil fuels, which will help stop global warming. He maintains that ordinary people can help save energy by purchasing energy-efficient light bulbs, appliances and automobiles. In addition, he claims that government and industry must do their part in fostering the use of clean energy sources and bolstering fuel efficiency. Bob Herbert is a columnist for the *New York Times*.

As you read, consider the following questions:

1. How much variance in energy consumption can there be between appliances with comparable features, according to Herbert?
2. According to the author, by how much can carbon dioxide emissions be cut when doubling the fuel economy of an automobile?
3. What clean energy sources does Herbert mention in this viewpoint?

So what can we do about global warming?

First, keep in mind the goal, which is to bring the potentially catastrophic warming under control by curtailing the release of carbon dioxide and other heat-trapping gases into the atmosphere.

What Ordinary People Can Do

Ordinary people can help immediately by becoming more energy efficient. Stop using the familiar incandescent light bulbs and replace them with compact fluorescent bulbs, which last much longer and use only a quarter of the energy consumed by conventional bulbs.

Compact fluorescent bulbs are significantly more expensive, but because they last so long (up to 10 times the life of a standard bulb) and use so little electricity, they are substantially cheaper in the long run.

Next, when shopping for an appliance—a refrigerator, a dishwasher, an air-conditioner—select the one with the highest energy efficiency rating. There will be a sticker on the appliance, telling you how much energy it uses. Pay attention. There can be a difference of 30 percent to 40 percent or more in the amount of energy consumed by appliances with comparable features.

Transportation Choices

Even more important is the choice you make in the car or truck you buy. Motor vehicles are responsible for about a third of the carbon dioxide emissions in the United States. The vehicles that are the most fuel efficient emit the least carbon dioxide. (Fuel economy and carbon dioxide emissions are inversely proportional. If you double fuel economy, you cut carbon dioxide emissions in half.)

According to the research and advocacy group Environmental Defense, if you buy a new car that gets 10 more miles per gallon than your old car, the amount of carbon dioxide reduction realized in one year will be about 2,500 pounds.

So buying a car or truck that suits your needs *and* is fuel efficient is a big help.

Honda and Toyota are bringing so-called hybrid vehicles onto the market in the U.S. Hybrids are cars that combine an internal combustion engine and a battery-powered electric motor. They are mid-sized cars that are achieving twice the fuel economy of conventional cars.

Dr. Paul Epstein, associate director of the Center for Health and the Global Environment at Harvard Medical School, summed the matter up as follows: "The issue is not so much what we are doing, but how we power what we are doing. That's the first step."

Move to Clean Energy Sources

Over the long term, the requirements are far more ambitious. Ideally, over the course of the next 100 or so years, a transformation will take place and most energy will end up coming not from fossil fuels like coal and oil, but from clean energy sources—the sun, the wind, hydrogen and non-polluting fuel cells.

"To get there at a cost that's affordable will require substantial technological development," said Dr. Michael Oppenheimer, the chief scientist of Environmental Defense.

Fooled into Conserving

Energy entrepreneurs and researches say that the widespread prosperity in the 1990's usually required conservation and efficiency to be dressed up as something else, like luxury. The Maytag Corporation came up with a superefficient front-loading washing machine several years ago called the Neptune that uses far less water and electricity than a conventional top-loader. But the company positioned it as a top-of-the-line European-style machine that was gentler on clothes and tacked on a price that was twice as high as its regular washers. Sales took off.

Kirk Johnson, *The New York Times*, October 21, 2000.

To move from our pollution-choked present to a future in which climate change is not a mortal threat will take more than that all-important first step of an enlightened citizenry buying cleaner cars and more efficient appliances. Tough action by Congress and the president is needed, and soon. And international cooperation, with enforceable agreements covering both industrialized and, ultimately, developing nations, will be crucial.

Among other things, the federal government can offer subsidies and other incentives to reduce the cost and foster the use of existing clean-energy technology, and to encourage the development of ever more efficient new technologies. And the government can—and should—develop more sophisticated strategies like emissions trading and more stringent requirements for reducing carbon dioxide emissions everywhere.

Global warming is the most serious problem we face in the 21st century. Last week an intensive analysis by a respected geologist at Texas A&M University suggested—as most scientists have been saying for some time now—that human activity, not natural factors, is the primary cause of the warming.

We caused the problem and we have within our grasp a variety of potential solutions. To ignore those solutions, to be aware of them but not make use of them, is not just profoundly destructive, it's suicidal.

| *"The truth is that energy conservation is virtually always a bust."*

Conservation Cannot Solve Energy Problems

Herbert Inhaber

Herbert Inhaber is a physicist and author of the book, *Why Energy Conservation Fails*. In the following viewpoint, Inhaber claims that conservation never succeeds in decreasing demand for energy because it ignores economic realities and human nature. He contends that demand for a commodity only decreases when shortages occur and prices rise, and since prices have remained low for fossil fuels—which remain abundant—demand naturally remains high. Furthermore, human nature guarantees that as some people conserve energy, other people will use up the energy that has been saved.

As you read, consider the following questions:
1. According to Inhaber, how many billions of dollars in savings has the United States lost as a result of its 1995 efficiency programs?
2. What happened to gasoline consumption after Congress mandated a doubling of gas mileage in 1973, according to the author?
3. According to Inhaber, how much did the price of a barrel of North Sea Brent crude oil drop in 1997?

A s usual, the sweltering summer heat has utilities calling on customers to conserve electricity. Meanwhile, President Bill Clinton's 1998 budget calls for a 20% increase in spending on energy conservation, for a total of $708 million. Mr. Clinton's budget writers say this investment will save consumers and businesses over $10 billion annually by 2005.

Energy Conservation Is a Bust

But a little perspective is in order. In describing its budget two years ago [1995], the Department of Energy said that its efficiency programs would save the nation $30 billion annually by 2005. Somehow $17 billion of those "savings" have already disappeared. The Energy Department is unable to supply any evidence for its latest claims.

The truth is that energy conservation is virtually always a bust. Consider the experience of England in the 19th century, when the coal mines seemed to be emptying. Then the father of quantitative economics, Stanley Jevons, observed that greater efficiency produces more energy use, not less. Jevons pointed out that Watt's steam engine was much more efficient than its predecessor, the Newcomen engine. Because Watt's engine was so efficient, demand soared. The engine ushered in the Age of Steam, and world coal use skyrocketed.

The lesson resonates today. Governments around the world continually trot out new schemes to reduce energy use and promote efficiency. Yet, as the Swedish economist Lennart Hjalmarsson notes, "I have not managed to find one single evaluation of energy conservation programs published in a scientific journal that shows the program has managed to reduce growth in electricity demand at a national or regional level and the program has been cost-effective."

The prime American example of this futility is government regulation of automobile gas mileage. Prompted by the Arab oil embargo of 1973, Congress mandated a doubling of gas mileage. What happened? Gasoline consumption rose from 1973 to the 1990s, as the roads were flooded with energy-efficient cars. Huge sport-utility vehicles crowd parking lots, also thanks to more efficient engines.

The ostensible reason behind the push for conservation is to put off the day when the last barrel of oil gushes out and the last lump of coal is hauled up. But if these resources were truly running out, we would not see their prices fluctuate and decline. If the premise behind conservation were true, prices would always rise. In fact, North Sea Brent crude oil has dropped $6 a barrel in 1997 alone.

But some environmental groups actually wish to hasten the day when we run out of energy. Greenpeace, which for years has demanded ever-greater government intervention to promote conservation, declared in Amsterdam on July 4: "We call for an immediate halt to exploration of coal and oil." Greenpeace had discussions with senior management of Royal Dutch/Shell; sad to say, no one captured on film the looks on the oil executives' faces.

Economics and Human Nature

Conservation fails because it takes no account of economics or human nature. Conservationists imagine the world as akin to a laboratory experiment with two engines, A and B. If we improve engine A's efficiency, the fuel saved can be devoted to running B longer.

The Performance Downside to Efficiency

Air conditioners rated as highly efficient under the federal standards are less reliable than average models and some do a poor job dehumidifying the air. "I've seen state of the art, high efficiency air conditioners in homes where there's mold and mildew on the walls," says Dave Debien, owner of Central City Air, in Houston. Many consumers are dissatisfied with the weak trickle from federally mandated low-flow shower heads. Some highly efficient refrigerators have freezer sections that don't stay as cold as they should to safely preserve foods for extended periods. Efficiency mandates often have a performance downside to go along with the energy savings.

Ben Lieberman, *Freeman*, April 1999.

I am not engine B. When I purchase a vehicle, I may go for a Lincoln rather than a Taurus. I might reason that the extra fuel cost compared to income wouldn't be that much, on the order of $3 a week. And if you trade in a BMW for a Kia, more gasoline is available for the rest of us. When a commodity becomes more plentiful, its price generally drops.

The combination of greater engine efficiency and rising disposable income has produced a true golden age of motoring. If the typical disposable income in the 1950s were entirely devoted to buying gasoline, the average person could drive about 75,000 miles a year. By the mid-1990s, that figure had risen to about 350,000 miles. That means it's entirely logical to buy Toyotas with V6 engines instead of the sewing machine-style four-cylinder engines of the 1970s, and to fill the driveway with vehicles that can't fit into the garage.

In the same way, what is "saved" by installing special light bulbs is often "wasted" on new hot tubs, exterior lighting and a host of other energy uses, as homeowners assume that their electric bills will drop off substantially.

In spite of these and dozens of other clear failures, the claims for conservation to solve virtually all our national energy dilemmas continue. Few if any are valid. While each of us can reduce energy use in one or two areas, we find that the nation gradually uses more. The government is called in to solve the problem. But "waste" is in the eye of the beholder: The government can no more outlaw it than it can mandate joy.

Periodical Bibliography

The following articles have been selected to supplement the diverse views presented in this chapter. Addresses are provided for periodicals not indexed in the *Readers' Guide to Periodical Literature*, the *Alternative Press Index*, the *Social Sciences Index*, or the *Index to Legal Periodicals and Books*.

Jonathan H. Adler	"The Problem with Wind Power," *Weekly Standard*, October 25, 1999.
Gary Benoit	"Energizing America," *New American*, March 27, 2000.
Robert L. Bradley Jr.	"The Growing Abundance of Fossil Fuels," *Freeman*, November 1999.
Robert L. Bradley Jr.	"Renewable Energy: Not Cheap, Not Green," *Policy Analysis*, August 27, 1997.
Ralph Cavanagh	"Resurrecting Energy Efficiency," *New York Times*, July 28, 1999.
Seth Dunn	"Power of Choice," *Worldwatch*, September/October 1997.
John Emshwiller	"California's Shortages Rekindle Its Efforts to Conserve Energy," *Wall Street Journal*, February 20, 2001.
Mark C. Fitzgerald	"Renewable Energy Today and Tomorrrow," *World & I*, March 1999.
Jeff Gersh	"Capitalism Goes Green?" *Amicus Journal*, Spring 1999.
Ben Lieberman	"Wasting Energy on Energy Efficiency," *Freeman*, April 1999.
Natural Resource Defense Council et al.	"Carbon Kingpins: The Changing Face of the Greenhouse Gas Industries," *Mulinational Monitor*, June 1999.
Fred L. Smith Jr.	"Prometheus Bound: The Case for Energy Use," *Regulation*, Winter 1998.
James Srodes	"No Oil Painting," *Spectator*, August 29, 1998.
Mitch Trachtenberg	"The Speech You Won't Hear at the Conventions: Solving Our Energy Problems," *Progressive Populist*, September 1, 2000.
Eric Weltman	"Here Comes the Sun," *In These Times*, February 7, 2000.

CHAPTER 2

Is Nuclear Power a Viable Energy Alternative?

Chapter Preface

The 1986 explosion of the Chernobyl nuclear reactor in the Ukraine is the worst nuclear accident on record. The explosion resulted in the immediate deaths of thirty-one people, and experts project that over 3,400 local residents will eventually die of cancer due to their exposure to radiation. At the Three Mile Island nuclear plant in Middletown, Pennsylvania, radioactive gas was released from the plant in March 1979. Although no deaths have been reported due to the accident, some researchers have documented a high incidence of stillbirths and of thyroid and lung cancer in people living in the region.

While many people point to these high-profile accidents as proof that nuclear power is dangerous, other analysts see them in a more positive light. Experts who support nuclear power claim that the accidents at Chernobyl and Three Mile Island have led to stricter regulations governing nuclear power plants. For example, in 1996, the Nuclear Regulatory Commission—the federal agency charged with monitoring the safety of nuclear power plants in the United States—began cracking down on power plants where safety problems had been reported. In addition, many nuclear power proponents argue that since the accidents, nuclear power plants have gotten safer. Douglas S. McGregor, director of Semiconductor Materials and Radiological Technologies, claims that the Three Mile Island accident "directly resulted in improved procedures, instrumentation, and safety systems."

However, those who oppose nuclear power point to the Chernobyl and Three Mile Island accidents as proof that nuclear power poses a serious health risk to people living near nuclear facilities. Opponents contend that no advanced technology can guarantee that such accidents will never happen again. Moreover, many critics claim that the regulatory agencies charged with overseeing the nuclear industry cannot be trusted. Rosalie Bertell, president of the International Institute of Concern for Public Health, maintains that the International Commission on Radiation Protection—an organization set up by the United Nations to promote the peaceful use of atomic energy—"exists largely to play down the effects of radiation on human health."

Reactions following the accidents at Chernobyl and Three Mile Island ranged from optimism that they would inspire safer operating procedures to anger that people were put in such danger. To be sure, public perception of a deepening energy crisis at the dawn of the 21st century has brought the debate about nuclear power once more to the fore. Authors in the following chapter debate whether or not nuclear power is a viable energy alternative.

| *"Energy planners around the world are discovering that the 'energy source of the future'—nuclear power—is rapidly becoming a thing of the past."*

Nuclear Power Is Not the Energy of the Future

Christopher Flavin and Nicholas Lenssen

Christopher Flavin is senior vice president at the Worldwatch Institute, an organization that monitors the environmental health of the planet. Nicholas Lenssen is a private energy analyst and a former senior researcher at the Worldwatch Institute. In the following viewpoint, Flavin and Lenssen argue that the world no longer views nuclear power as a viable energy alternative. The authors contend that nuclear accidents have caused the public to view nuclear power as dangerous. In fact, the authors claim that although Asia is continuing to develop nuclear power capabilities, the United States and Europe are shutting down their nuclear power plants and have cancelled plans to build new ones.

As you read, consider the following questions:
1. How much of the world's electricity do nuclear power plants generate today, according to the authors?
2. According to the authors, what percentage of French citizens thought that developing nuclear power should be their country's top energy priority?
3. According to Flavin and Lenssen, what two countries have the world's most active nuclear power construction programs?

From "Nuclear Power Nears Its Peak," by Christopher Flavin and Nicholas Lenssen, *World Watch*, July/August 1999. Copyright © 1999 by the World Watch Institute. Reprinted with permission.

A s the new century approaches, energy planners around the world are discovering that the "energy source of the future"—nuclear power—is rapidly becoming a thing of the past. Most countries have brought the construction of nuclear plants to a halt, and some are debating how fast to phase out the plants they now have.

From World's Fastest to Second Slowest

After growing more than 700 percent in the 1970s and 140 percent in the 1980s, the industry's total generating capacity has increased only 5 percent during the 1990s so far. In the last decade, nuclear has gone from being the world's fastest growing energy source to its second-slowest. In 1998, capacity fell by 230 megawatts. Wind power, by contrast, rose by 2,200 megawatts.

By the end of 1998, 429 nuclear reactors were operating worldwide—one fewer than five years earlier. All told, these plants generate 343,086 megawatts of power, or just under 17 percent of the world's electricity. Construction is taking place at 33 new reactors, but at least 14 of these are unlikely ever to be completed. Although global capacity will probably rise for another few years, it will almost certainly decline beginning in 2003 at the latest, as the closure of older, uneconomic reactors accelerates.

In Germany and Sweden, the political debate is not about whether, but about how quickly, to shut down the existing nuclear plants. In Canada and the United States, safety problems and economic pressures are leading electric power companies to shut down some of their nuclear plants earlier than planned. And in the last year, even France—the most pro-nuclear European country—has adopted a moratorium on further construction.

Although modest additional nuclear capacity is likely in East Asia in the next few years, it will not be sufficient to prevent the global output from this industry from peaking at less than one-tenth of the 4.5 million megawatts that the U.N.'s International Atomic Energy Agency predicted for the year 2000 back in 1974.

This massive shortfall stems in part from the safety problems revealed by the nuclear accidents at Three Mile Island [in Pennsylvania] and Chernobyl [in the former Soviet Union]. Additional difficulties have stemmed from the world-wide failure to develop a safe, permanent means of disposing of long-lived nuclear waste. As a result, public opinion has turned strongly against nuclear power in most countries, making it difficult if not impossible to build additional plants.

The industry's biggest problems are economic. Costs rose dramatically in the 1970s and 1980s, as operators struggled to deal with unanticipated safety problems. In the United States, for example, the cost of a plant shot from $200 per kilowatt in 1972 to $3,000 per kilowatt in 1985. Meanwhile, since the early 1990s, the cost of other new power technologies has fallen dramatically—offering clean, economical alternatives to the atom.

The technology that dominates the global power market today is the combined-cycle natural gas plant, which uses combustion turbines similar to those of jet engines. It costs roughly $700 per kilowatt, less than one-third the cost of nuclear plants. As nations around the world move to turn their monopoly electric systems into private, competitive businesses, they are finding that the power industry has lost interest in the nuclear option, which is economically obsolete.

In the aftermath of the 1979 Three Mile Island accident, the first market to evaporate was in the United States, where no new plants have been ordered since the disaster, and where nuclear generating capacity is now lower than it was a decade ago. U.S. power companies have not only stopped building nuclear power plants; they have closed five reactors since 1996 that had become too expensive to operate. Meanwhile, seven of Canada's 21 reactors have been "laid up" and are unlikely to operate again.

Nuclear Power in the United States

For the U.S. sector, the worst may be yet to come. Wall Street analysts and research firms believe that as many as one-third of US reactors are vulnerable to being shut down over the next five years, virtually ensuring that the nuclear share of power will fall well below 20 percent. Investors are sending a clear message that nuclear power is too expensive to compete as electricity markets become more competitive.

Some U.S. power companies have attempted to sell off their reactors, though only two have found buyers so far—at rock bottom prices. The twin of the reactor that melted at Three Mile Island (TMI), for example, was sold in 1998 for $23 million. This is just 5 cents for every dollar of book value, or 1 percent of the replacement cost. The Pilgrim plant in Massachusetts was sold in 1999 for $80 million, of which $67 million was for fuel.

Nuclear Power in Europe

Western Europe stayed with its nuclear expansion plans a bit longer than the United States did. But since the day the 1986 explosion at Chernobyl sent a poisonous cloud of radioactive dust across Europe, only three new reactors have been started, and the region's nuclear industry stopped expanding by the second half of the 1990s. Public opinion is strongly opposed to nuclear power in most European countries, and both the power industry and governments have responded to those concerns.

In France, which produces more than three-fourths of its

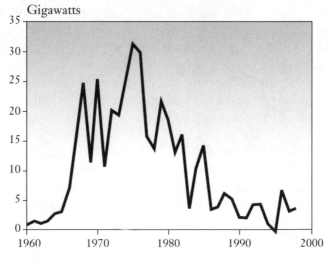

World Nuclear Reactor Construction Starts, 1960–98

Worldwatch

electricity from nuclear power, dramatic changes are afoot. A moratorium on additional nuclear plant construction was adopted by the French government last year, and the Environment Minister, Dominique Voynet, has called for making the ban permanent. A December 1998 poll found that only 7 percent of French citizens thought that developing nuclear power should be the top energy priority, compared to more than 60 percent who said the priority should be renewable energy. The state-owned utility, Electricite de France (EDF), which has in the past put virtually all its efforts into nuclear, has begun to invest in windpower.

In Germany, where the previous government shut down the nuclear power plants only in eastern Germany, the Social Democrat/Green government elected in October 1998 plans to phase out the 19 nuclear power plants in the western states that produce one-third of the country's power. The German power industry has vigorously opposed the government's plans, and the pace at which the country's reactors will be closed remains uncertain.

In Sweden, the coalition government first elected in 1994, on a platform that included plans to begin shutting down the country's reactors, is moving forward on its pledge. However, the owner of the first targeted reactor, Sydkraft, has taken the government to court and has delayed closure at least through this year. The Netherlands closed one of its two reactors in 1997, with the other one scheduled to be shut in 2004.

In the countries that make up the former East Bloc, including Eastern Europe, Russia, and Ukraine, nuclear power peaked a decade ago with the fall of the Berlin wall—and has since fallen by 10 percent. Some 68 reactors now operate, and only four more are under active construction. No new construction has started since economic reforms began. Although some of these countries hope to expand their nuclear industries, they face financial limitations and strong public opposition.

Nuclear Power in Asia

Asia remains the last stronghold for the nuclear industry, with 88 reactors operating and 26 under construction, although even here a slowdown is evident. Japan, which obtains 35 percent of its electricity from the atom, has only two reactors under construction, with work starting on one of them in 1999—the first new plant approved in ten years. Citizens in rural areas have strongly resisted construction of new plants, with some communities passing referenda prohibiting additional units.

A Colossal Failure

Many consumer-advocacy organizations . . . say nuclear power has proven to be a colossal failure. Nuclear power is more expensive and more technologically cumbersome to produce than was originally expected, they note. Citing worker errors and safety lapses at nuclear plants in recent years, they argue that producing nuclear energy remains a risky enterprise that ought to be abandoned for good.

In a similar vein, environmentalists argue that nuclear reactors pose a huge threat because they produce tons of deadly radioactive waste each year. Disposing of the waste, which may remain radioactive for 10,000 years or longer, is a challenge that the industry has yet to solve. No one, environmentalists add, can be certain of whether radioactive waste can be safely stored for that long.

Issues and Controversies on File, March 31, 2000.

South Korea, together with China, has the world's most active construction program, though the twin blows of political change and economic crisis have slowed down the pace and probably ensured that the nuclear share of electricity will plateau at the current level. The country's economic crisis depressed growth in electric power, and has driven up the costs of financing capital-intensive projects such as the six nuclear plants now being built. As in Japan, local communities have taken increasingly strong positions against building more reactors.

Taiwan is building two reactors, but these are widely regarded as the last two likely to be built in that country. Efforts to create commercial nuclear power industries in Indonesia, Thailand, and Vietnam have all failed in recent years. India launched construction on a new project (its first since 1990) last year, but the Hindu nationalist government fell in early

1999, leaving the country's nuclear plans in doubt. Today, nuclear energy only supplies 2 percent of India's power.

China, like South Korea, has six plants under construction, but its plans are more ambitious. China currently intends to go from the three reactors it operates today (supplying about 1 percent of the country's power) to 11 by 2006 and more than 50 by 2020. China's leaders are concerned about the country's heavy dependence on highly polluting coal, and its nuclear weapons establishment provides a strong base of political support for the industry.

China's ambitious program will likely fall short, however, as have those of other developing countries that have tried unsuccessfully to develop nuclear power in the past two decades. China faces foreign exchange limits, which will further constrain its nuclear ambitions, and early this year the government announced that it will delay additional plant starts. For the long run, China has a wealth of renewable energy sources, such as wind and solar power, that are likely to be the best alternatives to fossil fuels.

The same logic is likely to prevail at the global level. Although nuclear industry supporters argued at the 1998 climate negotiations in Buenos Aires that the problem of fossil fuel-induced climate change will necessitate a return to the atom, few governments—and even fewer private companies—seem to be buying the argument. Instead, they have responded to the climate change challenge by investing in new energy technologies such as solar energy and wind power. Although these new energy technologies provide tiny amounts of power today, they are already growing at the kind of double-digit rates that nuclear power enjoyed in the 1970s. And renewable energy promises to be immune from the kind of physical and economic meltdowns that crippled the nuclear power industry.

| "*Nuclear power, which provides about one-fifth of America's energy, is one of the most reliable and plentiful sources of electricity.*"

Nuclear Power Is the Energy of the Future

Patrice Hill

Patrice Hill contends in the following viewpoint that nuclear power has become less expensive than any other source of electricity. According to Hill, another significant benefit of nuclear energy is that nuclear power plants do not produce the carbon emissions—which are responsible for environmental degradation—that fossil-fuels plants do. Hill contends that political opposition to nuclear power has been waning as the United States faces an ever more threatening energy shortage. Patrice Hill is a reporter for the *Washington Times*.

As you read, consider the following questions:
1. According to Hill, how much of America's energy comes from nuclear power?
2. What new technology can guarantee the safety of nuclear power plants, according to the author?
3. According to Hill, why have new natural gas power plants failed to avert California's energy crisis?

From "Energy Crisis Rekindles Interest in Nuclear Power," by Patrice Hill, *Insight*, April 23, 2001. Copyright © 2001 by Insight Magazine. Reprinted with permission.

P erhaps the most visible sign that nuclear power is back: [High-tech region] Silicon Valley executives recently declared it the best solution to the chronic electricity shortage facing California. "Nuclear power is the answer," says Craig Barrett, chief executive of Intel Corp., "but it's not politically correct."

Reliable and Plentiful

Nuclear power, which provides about one-fifth of America's energy, is one of the most reliable and plentiful sources of electricity. Nuclear plants can run 24 hours a day, seven days a week and are not affected by drought or frigid weather like hydroelectric and other conventional sources.

But as Barrett acknowledges, resistance to nuclear power remains strong, particularly in Northern California, where the Green Party and other environmental groups are major political forces. Local officials consistently have blocked efforts to build new power facilities in the valley, one reason Intel says it won't expand there. Scott McNealy, chief executive of Sun Microsystems Inc., agrees.

"In terms of environmental and cost and competitiveness and all of the rest of it, I just don't see any other solution," the software executive said [in 2001] during a speech at the National Press Club in Washington, alluding to another nuclear selling point: It largely is pollutant-free and requires no disruptive drilling in sensitive environmental areas, unlike oil and gas exploration.

Cheap Power

Such endorsements from high-tech executives may be symbolic, but hard statistics indicate that nuclear no longer is a dying industry biding time as aging power plants deteriorate toward their inevitable burial. Today, with the cost of natural gas and oil soaring, mothballed nuclear plants considered too expensive to maintain and operate are being revived. The nation's 103 operating plants are producing a record amount of power, up 3.7 percent to 755 billion kilowatt hours last year, according to the Nuclear Energy Institute. About 90 percent of these are expected to seek 20-year extensions of their licenses.

Indeed, for the first time in more than a decade, nuclear production is less expensive than any other source of electricity generation. Improvements in maintenance procedures have meant less down time for refueling, enabling the plants to operate at a record 89.6 percent of capacity in 2000.

"It's the best year ever in performance," says Alfred C. Tollison, executive vice president of the Institute of Nuclear Power Operations. "The foundation is being put in place for a renaissance in nuclear power," though he adds, "that depends on the industry remaining accident-free."

Political Opposition

All sides agree that public perceptions about nuclear safety and nuclear waste remain significant obstacles. Because of that, no new plants have been built in the United States in the last two decades, and none are on the drawing board.

But there are signs that political opposition may not be as potent as in the past. The industry's clean record on safety after decades of operating power plants in the United States, France, Japan and other industrialized nations is vindicating nuclear's reputation, argue observers. Meanwhile, a new generation of technology is being developed that virtually could guarantee safety through automatic shutdown mechanisms designed to prevent even the remote possibility of a meltdown.

Exelon Corp. wants to start building a new plant—expected to be smaller, quicker and cheaper—using this new technology in South Africa by 2002 with the intention of exporting the technology to the United States. "Nuclear power is much safer than fossil-fuel systems in terms of industrial accidents, environmental damage, health effects and long-term risk," said Richard Rhodes, a Pulitzer Prize-winning journalist and author on energy issues, during testimony before the House Energy and Commerce Committee. "The U.S. nuclear-power industry has an extraordinary record of safe operation across the past 40 years, and I would submit to you that disposal of civilian nuclear waste is a political, not a technical, problem."

The Energy Department could designate a permanent disposal site—most likely Yucca Mountain in Nevada—as early as [2001] under procedures Congress established in 1987 that require extensive scientific review for safety. In fact, legislators quietly are taking renewed interest in nuclear power, with Republicans and some centrist Democrats advocating the energy source as a way to solve the country's energy crisis. Further chronic power shortages are expected in California [in the summer of 2001] and could crop up in New York and other Northeast cities in coming months as well.

Natural Gas Woes

During the 1990s, most utilities expanded power generation by building small, inexpensive units fired by natural gas, which became the power source of choice for environmental as well as economic reasons. Now, with the quadrupling of natural-gas prices in the last year, those gas-fired plants are expensive to run and are a major reason that wholesale electricity rates skyrocketed in California, bankrupting the state's utilities.

The woes faced by gas-fired plants, many of which are just coming on-line, will continue, energy analysts say. They predict that robust demand for gas from both power plants and home owners will keep prices elevated at around $5 per million British thermal units—double what they were at the end of 1999. Such prices make nuclear plants more expensive to build but cheaper to operate and competitive for the first time in years, say industry officials.

Support from the Left

With most opposition to nuclear power coming from the left, it's particularly significant that centrist Democrats are reconsidering it as an effective way to curb carbon-dioxide emissions, thought to cause global warming. Unlike coal-, natural-gas- and oil-fired plants, nuclear plants are free not only of carbon emissions but also sulfur dioxide, mercury and nitrogen oxide—noxious byproducts that have made fossil-fuel plants the biggest sources of air pollution in the United States. By contrast, nuclear plants provided about half of the total carbon reductions achieved by U.S. industry under a federal voluntary reporting program in 1999.

The Clinton administration gave nuclear a little-noticed

boost as it sought to find economical and relatively pain-free ways to comply with the steep cuts in carbon-dioxide emissions called for under the global-warming treaty. In negotiations over the treaty at The Hague in November 2000, U.S. negotiators waged a monumental fight with environmentalists and the 15-nation European Union over the use of nuclear power to curb carbon-dioxide emissions in developing countries.

"Nuclear power, designed well, regulated properly, cared for meticulously, has a place in the world's energy supply," said former vice president Al Gore at the Chernobyl museum in Kiev in 1998. Gore's running mate for president, Sen. Joseph I. Lieberman, D-Conn., also endorsed nuclear energy as "part of the solution to solving the world's energy, environment and global-warming problems."

Practical and Safe

Despite the rhetoric against it, nuclear energy has demonstrated its practicality, economic viability and safety. No one has been hurt by the operation of a nuclear plant built to U.S. standards. The Three Mile Island nuclear plant accident in Pennsylvania, despite its notoriety, produced a rise in nuclear radiation at the plant perimeter that was only a fraction of the yearly natural nuclear radiation. The key barriers to nuclear plant construction are not technical but institutional and political.

Bertram Wolfe, *San Francisco Chronicle*, December 8, 1998.

And Sen. Bob Graham, D-Fla., concerned about potentially catastrophic floods caused by global warming in his state, has spoken about his interest in nuclear energy's potential to reduce carbon emissions.

France, Japan and several other industrialized countries rely heavily on nuclear power to reduce their carbon emissions. A Nuclear Regulatory Commission study found that if the United States used nuclear power to the extent that France does—80 percent—it could in one fell swoop achieve the goals of the environmental treaty, which calls for a 10 percent reduction of U.S. emissions below 1990 levels. Also, nuclear power does not require the destructive drilling offshore and in the Arctic National Wildlife Refuge (ANWR) that would be required to produce significantly more oil and gas in the United States.

New Legislation Favors Nuclear Energy

"Nuclear power is not a magic bullet, but it should also not be a poison pill," said Graham who, like many Democrats, opposes drilling in the ANWR as well as in the Gulf of Mexico off Florida. "The technology exists to make nuclear power—already one of our cleanest energy sources—also one of our safest, most reliable and least expensive."

Graham is the cosponsor of a bill, introduced in March by Sen. Pete V. Domenici, R-N.M, to expand the use of nuclear energy and support advanced research into technologies to minimize nuclear waste. Two other Southern Democrats have signed onto that legislation—Sens. Mary Landrieu of Louisiana and Blanche Lincoln of Arkansas, as well as a raft of Republicans. "We'll be talking about this in 18 months," predicted Domenici. "The U.S. can't just sit by and say we don't need this. We need it."

The Senate's energy-development bill, introduced in March by Senate Energy and Natural Resources Committee Chairman Frank H. Murkowski of Alaska, also offers incentives for nuclear-power production, including liability protection in case of nuclear accidents. The industry also is expected to get support from the Bush administration, which views nuclear power as "an integral part of U.S. energy security," though it has not offered any detailed proposals. Recommendations from a White House energy task force headed by Vice President Dick Cheney are expected within weeks.

House Republican leaders also see nuclear power as a key component of a strategy aimed at enhancing national security through energy independence. They, too, are promising incentives for nuclear power in the House's energy bill later this year.

"The nuclear industry has been stagnant for years, yet it offers the capacity for clean and emissions-free power," says Rep. J.C. Watts Jr., R-Okla., and chairman of the House Republican Conference.

Environmental groups dispute the nuclear-industry's claim to be emissions-free and question whether it will remain competitive for long. Kit Kennedy of the Natural Resources Defense Council says extensive drilling will force natural-gas prices down again within a few years, making nuclear alternatives less attractive.

"We think natural gas will continue to be a lot more tempting than taking on the huge task of building new nuclear plants," which face stringent opposition from local activists, says Kennedy. Her group has challenged advertisements by the Nuclear Energy Institute that portray nuclear as "clean and green," asking both the Federal Trade Commission and the Better Business Bureau to investigate the claims. Neither agency has taken enforcement action.

"The nuclear age has now produced . . . one serious accident. Meanwhile, the production and consumption of fossil fuel yields a constant flow of accidents and disease, in addition to greenhouse gas."

Nuclear Power Is Safe

John Ritch

John Ritch is U.S. ambassador to the United Nations Organization in Vienna, which includes the International Atomic Energy Agency. In the following viewpoint, Ritch maintains that nuclear energy is safer than energy derived from the combustion of fossil fuels. He claims that unlike energy from fossil fuels, nuclear power is safe for the environment because it does not produce greenhouse gases or other pollutants. Furthermore, nuclear power plants generate relatively small amounts of toxic waste in comparison with the production of waste by fossil fuel plants, he asserts.

As you read, consider the following questions:

1. What does Ritch blame for the accident at the Chernobyl nuclear power plant?
2. What is the sole large safety problem involving nuclear power, in the author's opinion?
3. According to Ritch, how much nuclear waste comes from industrial and medical activities as opposed to power production?

One myth, which exercises a powerful hold on the public mind, is that a nuclear power . . . constitutes a kind of bomb—likely, in case of accident, to explode or to release massively fatal doses of radiation. This myth is embodied in collective memory by the accidents at Three Mile Island [in Pennsylvania] and Chernobyl [in the former Soviet Union]. The power of those two images far exceeds what is warranted by the facts.

Three Mile Island and Chernobyl

At Three Mile Island in 1979, the simple truth is that public health was not endangered. Despite a series of mistakes which seriously damaged the reactor, the only outside effect was an inconsequential release of radiation—negligible when compared to natural radiation in the atmosphere. The citizens of the Three Mile Island area would have received more radiation by taking a flight from New York to Miami or standing for a few minutes amid the granite of Grand Central Station. The protective barriers in the reactor's design worked.

By contrast, the accident at Chernobyl in 1986 was a tragedy with serious human and environmental consequences. Chernobyl was a classic product of the Soviet era. A gargantuan reactor lacked the safety technology, the procedures and the protective barriers considered normal elsewhere in the world. The fire led to a massive release of radiation through the open roof of the reactor. More than two dozen firemen died from direct radiation exposure.

A conference sponsored by the World Health Organization (WHO) on the disaster's tenth anniversary issued a report based on intensive study of the 1.1 million people most directly exposed to the fallout. The main finding was a sharp increase in thyroid cancer among children; 800 cases of the disease had been observed, from which three children had died, with several thousand more cases projected. The report also predicted 3,500 radiation-induced cancer deaths, mainly late in life.

These statistics do not minimize the gravity of what happened at Chernobyl, but they place that singular event in perspective. The nuclear age has now produced more than 8,000 reactor-years of operational time—and one serious accident. Meanwhile, the production and consumption of fossil fuel yields a constant flow of accidents and disease, in addition to greenhouse gases. In the years since Chernobyl, many thousands have died in the production of coal, oil and gas; and millions each year are afflicted with pollution-induced disease resulting from the use of carbon fuels to produce energy which could be produced by nuclear power. According to the WHO, 3 million people die each *year* due to air pollution from a global energy system dominated by fossil fuels.

Nuclear Safety Culture

The question is: what has been done to prevent another Chernobyl? While Chernobyl severely damaged the standing of nuclear power, it inspired important advances in the global industry. Just as [Iraq's president] Saddam Hussein helped to strengthen safeguards against [nuclear weapons] proliferators, Chernobyl accelerated the arrival of a stronger nuclear safety culture. National regulatory agencies, a new World Association of Nuclear Operators and the International Atomic Energy Agency (IAEA) work together to promulgate state-of-the-art knowledge. Two years ago, a Convention on Nuclear Safety introduced a system of peer review to detect any deviation from the high safety standards which are now the norm.

For the total of some 440 power reactors (half in Europe) operating in 31 countries, and producing 17 percent of the world's electricity, only one large safety problem remains: in three countries of the former Soviet empire some 15 plants of the Chernobyl type are still in use. Although now equipped with safety upgrades and better trained personnel, these reactors fall short of current standards and must be phased out as soon as alternative energy supplies can be funded and installed.

Elimination of Chernobyl-style reactors will be an important step in ensuring that the industry will only have reactors of the most modern design. Building on a large base of operating experience, today's reactors are engineered on the principle of "defense in depth," ensuring against a release into the environment even in the case of a severe internal accident. Moreover, designers believe that the newest plants would experience such an environmentally harmless event no more than once in every 100,000 reactor-years of operation. Advanced plants now under development will have even less risk of internal damage.

The Question of Nuclear Waste

The fact that modern reactors are immensely safe shifts attention to the question of nuclear waste. The myth is that, regardless of reactor safety, the resulting waste is an insoluble problem—a permanent and accumulating environmental hazard. The reality is that, of all energy forms capable of meeting the world's expanding needs, nuclear power yields the least and most easily managed waste.

Radiation Can Be Healthful

In 1991, . . . the National Cancer Institute published a report concluding that there is "no general increased risk of death from cancer for people living in 197 U.S. counties containing or closely adjacent to 62 nuclear facilities." Moreover, demographic studies have since revealed that cancer rates vary inversely with exposure to background radiation. People living on the Rocky Mountain Plateau receive the highest doses of background radiation in the country (through radioactive minerals in the mountains and greater exposure to cosmic rays) yet have the lowest rates of cancer in the country. This has spawned a counter-theory which says that high levels of background radiation may be *healthy*. Just as a vaccine stimulates the immune system against microbial invaders, so small doses of radiation may stimulate the body's known mechanisms for repairing genetic damage.

William Tucker, *Weekly Standard*, April 2, 2001.

The challenge of climate protection arises precisely because it is fossil fuel consumption, not nuclear power, which presents an insoluble waste problem. The problem has two aspects: the huge volume of waste products, primarily gases and particulates; and the method of disposal, which is dispersion into the atmosphere. Neither seems subject to amelioration through technology.

In contrast, nuclear waste is small in volume and subject to sound management. Most nuclear waste consists of relatively short-lived, low and intermediate level waste—annually, some 800 tons from an average reactor. Such waste can be handled safely through standard techniques of controlled burial or storage in near-surface facilities. Half of such waste comes from industrial and medical activities rather than from power production.

High level waste consists of spent fuel or the liquid waste which remains after spent fuel is reprocessed to recover uranium or plutonium for further use. The annual global volume of spent fuel from all reactors is 12,000 tons. This amount—tiny in comparison to the billions of tons of greenhouse gases and many thousands of tons of toxic pollutants being discharged annually—can be stored above or below ground. Moreover, the volume decreases considerably if the fuel is reprocessed. The 30 tons of spent fuel coming from the average reactor yield a volume of liquid waste of only 10 cubic meters per year.

Safe Storage Techniques

Even with twice today's number of reactors, the annual global volume of liquid waste, if spent fuel were reprocessed, would be only 9,000 cubic meters—the space occupied by a 2-meter high structure built on a soccer field. Liquid waste from reprocessing can be vitrified into a glass which is chemically stable and subject to a variety of remarkably safe storage techniques. Indeed, the use of those techniques in long-term storage is now more a political than a technical question.

So far, as a result of political obstacles, nations employ various methods of interim storage because no long-term disposal site has been licensed in any country. A number of countries, however, are developing repository concepts. Under consideration are deep underground geological formations such as solid salt domes and granite tunnels which are impervious to water and thus to the leaching of materials. If such sites were used, this protection would be compounded by a series of other barriers: the vitrified state of the waste, high-endurance storage canisters, and a surround of absorbent clayfill. According to the IAEA, even if these barriers were not used, "the long path through the host rock to the surface would probably ensure sufficient dilution so as to pose little risk to human health or the environment." Moreover, storage sites can be designed so that all material remains under strict supervision—and subject to retrieval in the event that technological advance offers new opportunities for retreatment.

Clearly, the management of nuclear waste must meet high standards not only of public safety but also of public acceptance. A first step requires a broader understanding of the waste issue not as a disqualifying liability of the nuclear industry but as a matter of momentous social decision. The choice is between the reckless dispersal of horrendous volumes of fossil fuel emissions and the careful containment of comparatively limited quantities of spent nuclear fuel. To give a stark example: if Europe today were to eliminate nuclear generated electricity and revert to traditional fossil fuel power, the extra greenhouse gases created would be the equivalent of doubling the number of cars on the road.

"The risks to us [from nuclear power], to our children and to life on earth—the economic, environmental and social costs—are wholly unacceptable."

Nuclear Power Is Dangerous

Chris Busby

Chris Busby is a physical chemist, an independent researcher on the effects of low-level radiation, and the author of the book, *Wings of Death: Nuclear Pollution and Human Health*. Busby argues in the following viewpoint that pollution caused by nuclear weapons production and nuclear power generation presents serious environmental and human health risks. For example, he claims that the incidence of childhood leukemia is several times higher than average in areas surrounding nuclear facilities. Busby asserts that radioactive fission products cause genetic damage that leads to all types of cancer and other illnesses such as heart disease.

As you read, consider the following questions:
1. How much higher than the national average was the incidence of childhood leukemia in West Berkshire, according to Busby?
2. According to the author, how does the Harwell nuclear facility dispose of its radioactive waste?
3. What was the risk of dying from leukemia in the years between 1981 and 1995 for children up to fourteen years old, according to Busby?

Human psychology is well adapted to evaluating risk. We do not do this probabilistically, but on the basis of best and worst case: thus we are prepared to spend money on a lottery ticket and chance winning a huge sum, even though we are well aware of the vanishingly small likelihood there is of our winning anything at all. By the same token, we will not take a risk, however infinitesimally small, if the outcome is death. People are not impressed by arguments showing that the risk of fatal cancer per milliSievert dose of radiation exposure is five deaths per hundred thousand persons exposed even if such a number were guaranteed correct. The fact is, five people will die and one of them may be you! Would you allow your child to play in the park if there were a poisonous snake loose somewhere within it, however tiny the probability of being bitten? What if the snake were invisible, and its bite caused cancer or leukemia?

Invisible Snakes

This is a lesson that the nuclear establishment apologists should have learned by now. Not that we were ever given the option of deciding. The invisible snakes were released from their wicker baskets in 1945 when the A-bomb was dropped [by the United States] on Hiroshima, [Japan] and they have been released ever since, from nuclear sites all over the world, from atmospheric weapons testing, from [nuclear power plant] accidents like Windscale, Three Mile Island and Chernobyl.

"What does this have to do with us?" says Middle England, sitting outside at the 'Traveller's Rest.' Sweet Thames flow softly while I sing my song.

In 1983, the Sellafield leukemia cluster was first reported. Children were ten times more likely to contract the disease near Europe's largest nuclear reprocessing plant [which uses spent nuclear fuel for energy]. Attention was focused on the other reprocessing plants in Europe; Dounreay in Scotland and La Hague in France. They turned out to have increased risk of eight and fifteen times the national averages respectively. Other nuclear sites in England and Wales were examined. West Berkshire, which contains the Atomic Weapons Research Establishment at Aldermaston and also the Royal

Ordnance Factory, at Burghfield, also turned out to have a modestly elevated increase in childhood leukemia, about 40 per cent higher than the national average.

All this was rather alarming, of course, and so the government set up a new, and nominally independent, committee to examine the possibility of a link between the radiation from the nuclear sites and the increases in the disease, a screamingly obvious association to the ordinary person. "No," said the independent Committee on Medical Aspects of Radiation in the Environment (COMARE) "radiation cannot be the cause because the levels are too low." "Hurrah!" cried Middle England and breathed a sigh. Interestingly, however, the address of this committee is c/o The National Radiological Protection Board, Chilton, Didcot, Oxon. And anyone who has visited the site can tell you that Chilton, Didcot, Oxon, is also Harwell, the first and foremost site of the United Kingdom Atomic Energy Authority, and one of the first sources of radioactive man-made pollution in the [United Kingdom] (UK).

Radioactive Rivers

Harwell, like Sellafield, has been releasing radioactive isotopes to the environment since 1948. However, unlike Sellafield, there is no Irish Sea for the radioactive isotopes to be dispersed or diluted into. They were released as gases to the air, via a sewer to a local stream (the Lydebank brook which passes through the village of East Hendred), and via a pipeline to the river Thames at Sutton Courtenay. Half a million tons of radioactive waste each year is pumped through this pipe. In the summer of 1980 when I explored the Thames in my motor barge *Nidd*, I remember mooring alongside a pipe mouth at Sutton Courtenay. I have a photograph of my daughters, then aged 14 and 15, diving into the river there; the sun shines brightly in the photograph. They are frozen in space together above their reflections, between the bank and the water surface. On my desk is a report of measurements made by Harwell scientists of radioactivity in the soil at Sutton Courtenay waterworks in 1996 showing levels of 4,800 Bequerels of radioactivity per kilogram of soil. This makes the mud considerably more radioactive than the 400 Bq/kg which defines low-level nuclear waste under the Radioactive Substances Act 1993 and should mean that it has to be disposed of in a licensed site like Drigg in Cumbria. There are 4,800 radioactive disintegrations per second taking place per kilogram of this soil, 4,800 clicks on a Geiger counter per second, 4,800 primary energetic particles capable of smashing through living tissue, killing cells and mutating cells, ultimately, perhaps, to cause cancer. In the river, 100 meters downstream of the pipe, there are sediment Plutonium levels at 80 Bq/kg. A mile downstream the figure is 32 Bq/kg.

© Clay Bennett. Reprinted with special permission of King Features Syndicate.

The level for Plutonium should normally be zero, but, because of the fallout from atmospheric testing in the 1960s it is about 0.05 to 0.2 Bq/kg. Within a mile of the pipe end, National Power abstracts this radioactive water for Didcot Power Station and cycles the radioactive particles back into the atmosphere all over South Oxfordshire.

Nuclear Cocktails

Drinking water for many local towns and villages is taken from the Thames, whose water is drunk in London too. Recent declassification of documents from the Public Records Office made it clear that the Metropolitan Water Board regarded with "great regret and dissatisfaction" the original government decisions to permit the water to be contaminated by Harwell and Aldermaston with a nuclear cocktail of radioisotopes. Dr Charles Hill, for the Ministry of Housing, argued that there would be genetic disorders. He was overruled.

The river Kennet in Berkshire is similarly affected by releases from the Atomic Weapons Establishment at Aldermaston (which also has a long pipe to the Thames at Pangbourne) and the Royal Ordnance Factory, at Burghfield. Aerial discharges from all three sites can be inhaled by local people, can fall on crops, become incorporated into food and/or be washed into the rivers. If we consider the 800 square kilometer triangular area between Oxford, Newbury and Reading as a sink for all this radioactivity, and compare the density of pollution, in Bequerels per square meter, with the pollution to the Irish Sea from Sellafield, we find that the latter pales in significance. Where can it go to? Will the Thames and Kennet wash it all away: through Wallingford, Goring, Reading, Henley, Marlow, Maidenhead, Windsor, Chertsey, Kingston, Richmond, Chiswick and through the Pool of London to the sea? Will this journey dilute it to a safe level? Is there a safe level?

Cancer Statistics

One thing we can do to try to find out is to look at the cancer statistics for the areas involved, particularly childhood leukemia. Our initial attempts to obtain leukemia-incidence figures were consistently blocked by the Oxford Cancer Registry which opposed our access to the data. Perhaps this had something to do with the fact that the head of Oxfordshire Health Authority, Dr. Peter Iredale, was, before his appointment in 1992, the Director of the Harwell laboratory. We had to buy mortality data from the Office of National Statistics and we used this to analyze deaths from child leukemia in the areas of the county districts near the Thames and Kennet valleys above Reading. For the fifteen years between 1981 and 1995 the risk of dying from leukemia in the 0–14 age group was 2.5 times the national average in the South Oxfordshire Thames valley. For the Kennet valley it was 1.9 times the national average. Of the seven county districts in the region, Oxford city, Cherwell, West Oxford, Vale of the White Horse, Reading, South Oxfordshire and Newbury, it was only the last two, those that received the radioactive discharges, that had significantly increased childhood leukemia. We reported this result, together with measurements of radioisotopes recently made in the area, in the *British Medical Journal* in August [1997], causing consternation and a great deal of media interest.

Nuclear Plants as Nuclear Weapons

While nuclear energy is economically attractive, its power plants carry a danger that is seldom discussed. They are vulnerable to an enemy attack. Nuclear reactors are latent nuclear weapons, hostages to a potential enemy who could threaten to devastate them. Besides dispersing deadly radioactive material all over the landscape, a successful attack would severely cripple the target's electrical production. More nuclear power plants would only create more hostages for a potential enemy, dramatically raising the stakes of war for nations with nuclear reactors.

Center for Defense Information, *Defense Monitor*, vol. 29, no. 1, 2000.

In our letter we repeated our plea for a re-evaluation of the health effects of chronic exposure to low-level radiation from man-made radioactive pollution. These fission radioisotopes, Strontium-90, Caesium-137, Iodine-131, Tritium, Plutonium and so forth are substances that never existed on Earth before 1945. They mimic the very elements that life has evolved to use within cells. Yet they are unstable and radioactive. How, we ask COMARE, can they be compared with the effects of X-rays or natural background radiation? It is like comparing the energy absorbed as you sit and warm yourself in front of a fire with that transferred if you were to eat a hot coal. COMARE are not impressed. They continue, like the law, to use the risk factors for radiation-induced leukemia and cancer derived from the large single external exposure of the Japanese at Hiroshima.

The genetic damage caused by exposure to ingested and inhaled radioactive fission products does not only manifest itself as leukemia in children. It is a cause of all cancers and many other illnesses including breast cancer and heart disease. Weapons fallout is arguably the origin of the present cancer epidemic.

Outlaw Nuclear Pollution

I have been sent a cutting from the *Didcot Herald*. It gives an account of the funeral of a young man of 31 who has just died of cancer. He was clever, enjoyed rock music, he worked at Harwell, went on to get a degree, but was made redundant. I wonder if he was bitten by an invisible snake. I wonder if my daughters were, or if I was, back in 1980 at Sutton Courtenay on the Thames? There were no notices. I had no Geiger counter. How could I have known what the pipe was?

The nuclear project began in the atmosphere of the Cold War. Perhaps we knew less then about the effects of low-level radiation than we do now. Now we no longer need nuclear weapons. Who would we drop them on? Nuclear power, it is now clear, is by no means "too cheap to be metered." The risks to us, to our children and to life on earth—the economic, environmental and social costs—are wholly unacceptable. It is time to outlaw nuclear pollution. We are right to be frightened of invisible snakes.

> *"If we are to have reliable and affordable electricity . . . the time to move ahead with construction of new nuclear plants is now."*

Nuclear Energy Is Affordable

Mary L. Walker

In the following viewpoint, Mary L. Walker asserts that nuclear power plants generate electricity more inexpensively than do plants burning fossil fuels. She claims that nuclear power can be generated so cheaply that it more than pays for the safe disposal of the radioactive wastes produced by its generation. In addition, Walker contends that nuclear power is less expensive in the long run than electricity from fossil fuels because it generates none of the environmental costs that the combustion of oil, coal, and gas do. Formerly the assistant secretary for environment, safety, and health at the U.S. Department of Energy under the Reagan Administration, Mary L. Walker works as an environmental lawyer.

As you read, consider the following questions:
1. How many cents per kilowatt-hour does nuclear generation cost compared with the cost of generating coal, oil, and natural gas, according to Walker?
2. According to the author, what is one good economic practice that nuclear plants have used as a way to save money?
3. How many megawatts of electricity did the now-closed Rancho Seco nuclear plant produce, according to Walker?

Just a decade ago, few people thought there was a future for nuclear power in the United States. We read constantly about the trials and tribulations of the nuclear power industry. Pundits argued that competitive cost pressures, rather than the old concerns about safety and nuclear waste, would bring the nuclear industry to its knees.

Paying Its Way

That hasn't happened. Today, California's two nuclear facilities—Diablo Canyon and San Onofre—are the state's two biggest power sources, generating more than 4,000 megawatts between them. And they are among the least expensive sources of power in the state, providing electricity at less cost than even the newest power plants that burn natural gas.

Far from being an atoms-for-peace relic heading for extinction, nuclear power now sets the competitive bench mark for electricity generation. Production costs at nuclear plants throughout the country have dropped dramatically in the past few years, and efficiency today is at an all-time high, resulting in generating costs that average 1.83 cents per kilowatt-hour.

The cost of generating a kilowatt-hour from coal is higher at 2.07 cents. Oil comes in a distant third at 3.18 cents per kwh, while using natural gas to generate electricity costs 3.52 cents per kwh. Significantly, the lowest cost nuclear plants—which provide electricity for as little as 1 cent per kwh—generally boast the highest levels of safety, performance and reliability.

Because competition provides an incentive to operate nuclear plants more efficiently, utilities are measuring their plant performance against the best reactors here and abroad. Utilities also have created mechanisms to share good economic practices, much as they share good safety practices. For example, large cost savings have been achieved by shortening the time it takes to refuel nuclear plants and by reducing unplanned shutdowns.

Besides, nuclear power has from the outset paid its full costs in protecting the environment, from paying the expenses of its own regulation to contributing to a federal fund to dispose of nuclear wastes.

Later in 2001, the Department of Energy (DOE) is scheduled to announce the results of its decade-long scientific study of Yucca Mountain in Nevada, the designated site for the permanent disposal of spent nuclear fuel currently being stored at nuclear plants around the country. If DOE determines that Yucca Mountain is a suitable site for the repository—and there has been nothing said to suggest otherwise—President George W. Bush Jr. could announce a timetable for its opening.

Nuclear Efficiencies

Over the past 20 years, nuclear plant efficiency has risen sharply. It used to be that plants were shut down nearly as much as they were running, according to the Nuclear Regulatory Commission (NRC). Now, they routinely operate at 80% capacity. Refueling shutdowns that used to last several months have been accomplished in as little as three weeks.

The new efficiencies helped make nuclear energy a low-cost alternative to fossil fuel plants. John Reed, president of Navigant Capital, a firm that advises utilities selling power plants, said that as long as natural gas prices stay somewhat above historic lows, nuclear plants will have a cost advantage.

Terry McDermott, *Los Angeles Times*, February 9, 2001.

Consequently, there is recognition within the financial community that nuclear plants are increasingly attractive assets in a deregulated electricity market. Recently, quite a few nuclear plants have become objects of bidding wars as a result of an improvement in their operations.

This has led to a consolidation in the nuclear industry with several utilities vigorously competing to buy plants because nuclear-produced electricity is less costly to deliver than power from plants burning coal, oil or natural gas.

Environmentally Friendly

In addition to its competitive production costs, nuclear power possesses other important advantages. Of all the ways to generate electricity, nuclear is the most environmentally benign since there are no greenhouse gases or particulates such as sulfur dioxide or nitrogen oxides released into the air.

Yes, hydropower and other renewable energy sources can

make the same claim, but the construction of dams and lakes often does damage to natural as well as human habitats. And wishful thinking about solar energy and wind power should not be allowed to dominate our thinking to the extent that we fail to develop those sources of energy that can make a decisive contribution.

But will we learn from our mistakes? Conservation and energy efficiency, though essential, cannot provide the electricity needed for our digital economy. Not every proposal for a new power plant need be approved, but there are some avenues that should be avoided.

Californians might have been spared rolling blackouts if Sacramento had not forced the closure of the Rancho Seco nuclear plant in 1989, with the loss of the unit's 913 megawatts. Nor should we ignore the forced shutdown of the Trojan nuclear plant, a 1,100-megawatt unit, in Oregon. Within the last decade, Washington State scrapped four nuclear plants under construction that would have provided more than 4,000 megawatts of electricity. At least some of that power would have been made available to California.

What is indisputable is that the West lags behind every other region in the use of nuclear power, which makes us far more susceptible to sudden spikes in fossil fuel prices. If California is going to meet its energy requirements while reducing the emission of greenhouse gases and other pollutants, federal and state regulators will need to make it easier to site and construct nuclear plants. At the same time, our nation needs to move ahead aggressively with construction of a nuclear waste repository at Yucca Mountain in Nevada.

As Pulitzer Prize-winning historian Richard Rhodes told the House science subcommittee on energy and the environment last summer: "Nuclear power is demonstrably the greenest form of large-scale energy generation at hand. . . . The fundamental advantage of nuclear power is its ability to wrest enormous energy from a small volume of fuel."

If we are to have reliable and affordable electricity from power plants that don't foul the air or warm the planet, the time to move ahead with construction of new nuclear plants is now. Nuclear power should be part of the answer to the energy question.

| *"Nuclear power 'plants have generated some of the most expensive power in the country.'"*

Nuclear Energy Is Costly

Laura Maggi

In the following viewpoint, Laura Maggi argues that electricity generated by nuclear power has always been more expensive than energy generated from the combustion of fossil fuels. Furthermore, she claims that energy deregulation has made nuclear energy even more costly. According to Maggi, in order to compete with other companies that are not saddled with costly nuclear power plants, utilities have begun to charge consumers for costs associated with the construction and operation of their expensive nuclear facilities. Laura Maggi writes for *American Prospect* magazine.

As you read, consider the following questions:
1. What are "stranded costs," according to Maggi?
2. According to the author, what is the real cost per kilowatt-hour of nuclear power generation if capital costs are figured in?
3. How much have consumers already paid into the decommission fund for Entergy's Pilgrim plant, according to Maggi?

Excerpted from "Making White Elephants Fly," by Laura Maggi, *American Prospect*, volume 11, number 8, February 28, 2000. Copyright © 2000 by The American Prospect, 5 Broad Street, Boston, MA 02109. All rights reserved. Reprinted with permission.

In the summer of 1998, after about a year of peddling its Oyster Creek nuclear plant and finding no takers, GPU Inc. appeared resigned to shutting the unit down. Aging, inefficient, and economically uncompetitive, Oyster Creek was a prime example of how nuclear power—the ultimate energy boondoggle—wouldn't survive in the new world of deregulated energy markets. But this past fall, GPU announced it had actually found a buyer. At $10 million, one-sixtieth of the plant's $600-million valuation, AmerGen Energy Company got a real bargain. In fact, the company, a joint venture between Philadelphia-based PECO Energy Company and British Energy, was formed explicitly to scavenge the nation's unwanted nuclear units. It has made similar offers on six other plants and plans to buy many more. PECO, which is merging with the Midwest's utility behemoth Unicom, is fast becoming the country's nuclear powerhouse. Another company, Entergy Corporation, is also eyeing the nuclear market and has already completed a deal to buy a plant in Massachusetts.

White Elephants

Given the industry's history, expanding into nuclear power might seem ridiculous. But as Oyster Creek's $10-million price tag demonstrates, nuclear reactors are going for clearance prices—subsidized, of course, by consumers. Thanks to these subsidies, white-elephant nuclear power plants are enjoying a second childhood. Ironically, these bailouts are part of energy "deregulation," which supposedly subjects electric utilities to the discipline of market forces.

In the words of anti-nuclear activist Paul Gunter, "These plants are selling for the cost of fuel, with the reactors tossed in as a freebie." The utilities, eager to shed their nuclear holdings, can afford to part with the plants for cheap because under state deregulation laws, the consumer pays off the remaining construction debt—the enormous capital costs that made nuclear power so expensive and, therefore, uncompetitive.

The Great Bailout

So far, 24 states have begun opening their energy markets to competition. These are states with higher-than-average energy prices, usually because of utilities' unwise investments. In the past, nuclear power plants sometimes ate up billions of dollars before producing a single kilowatt. The escalating costs of building nukes have been ascribed both to tougher regulations implemented after [Pennsylvania's] Three Mile Island accident in 1979 and to utility incompetence. These plants have generated some of the most expensive power in the country.

During the past few years, as states moved to restructure their energy markets, legislators debated how to deal with nuclear plants and other bad investments, which the utilities euphemistically termed "stranded costs" or "stranded assets." The utilities contended that they would not be able to compete in an open market against firms with no previous baggage unless consumers paid off the remaining capital costs.

In the era of regulation, customers typically paid for bad decisions that had been sanctioned by regulators. In the jousting over the terms of deregulation, some consumer and environmental groups argued that utilities often invested in these ventures over local objections, so utility shareholders should bear some, if not all, of the costs as the companies move to a competitive market. However, in every state deregulation scheme, utilities and their shareholders won huge bailouts by ratepayers, amounting to billions of dollars that will be tacked onto millions of customers' electricity bills over the course of several years.

Footloose and Debt Free

Sloughing off the debt makes nuclear power plants artificially competitive, even against the cheapest power sources. Because most plants have two or three decades left on their licenses, this could amount to a very shrewd investment for AmerGen or Entergy. Evaluating only production costs, the Nuclear Energy Institute (NEI) predicts plants' generating prices will be roughly on par with coal: nuclear power at 1.9 cents per kilowatt, coal at 1.8 cents per kilowatt. For example, the Limerick plant in Pennsylvania, which is owned by PECO, generates power at 1.43 cents a kilowatt-hour after deregulation, when most of the capital costs have been written off. But if capital costs are figured in, the company estimates the plant's power would cost eight cents a kilowatt-hour.

During atomic energy's half-century of commercial existence, the industry has survived on the public dole, clinging to life supported by subsidy after subsidy. Nuclear has become the most heavily subsidized energy source, from the industry's federally backed insurance limiting utilities' liability to ratepayers who absorbed the enormous construction costs through years of high energy prices, to billions of dollars in federal research and development funding. Now, even their recycling to new owners is being subsidized.

The Future of Nuclear Energy

The long-term health of nuclear energy does not appear robust. Given the huge start-up costs associated with building and maintaining a reactor, even nuclear advocates admit that it is unlikely that utilities will build new plants any time soon, especially in a deregulated power environment. Meanwhile, the country's fleet of reactors is aging. Most nuclear power plants were built with the expectation that they would be able to produce power for only 40 years. Consequently, unless the economic and environmental costs of coal and other fuels become high enough to revive interest in building new nuclear reactors, the industry will likely wither away over the next three decades.

Issues and Controversies on File, March 31, 2000.

But this latest ratepayer bailout changes the very economics of the nuclear business. This is particularly galling because one of the initial promises of energy deregulation was that the most efficient technologies would win out in a free market, dooming economic and environmental dinosaurs like nuclear. Instead, the mechanics of "deregulation" have produced a sheltered market in which uneconomical facilities are shielded from the rigors of competition and ratepayers are no longer protected by regulation. Plus, the new nuclear owners will most likely find a heightened demand for their product in a market characterized by increasingly tight capacity, where the strong economy and consumer needs are eating away at surpluses that have typified U.S. energy markets in recent years.

This is not to say that the new owners of a nuclear plant aren't making a financial bet. Major repairs on nuclear plants can be very expensive. More importantly, the sale of used plants does not herald a full-scale nuclear revival. Nuclear has been on the wane since the Three Mile Island accident and the vigorous protests that followed, when plans for nearly 100 plants were scrapped. Even if the Nuclear Regulatory Commission approves 40-year extensions of some plants' licenses, nobody thinks it makes economic sense to build a new plant.

Nuking Consumers

Strangely, these transactions are often presented as a good deal for consumers. In news reports about the $10-million sale of Oyster Creek, GPU spokesmen have highlighted the $200 million ratepayers will supposedly save. But those savings are a calculation of the amount consumers would have paid into the decommissioning fund if the plant were shut down early, leaving out the roughly $590 million ratepayers will cough up to relieve GPU of the debt associated with the plant.

Or take the example of the Pilgrim plant in Massachusetts. Entergy, a New Orleans-based company that is also the public utility in Louisiana, paid Boston Edison Company $81 million for a plant with a book value of $700 million. Under the agreement, approved in the summer of 1999, Entergy also received a decommissioning fund worth $466 million—a dowry that will pay to dismantle the plant once the license expires and the reactor is shut down for good. Consumers have already paid around $200 million into the fund, and Boston Edison will raise the rest from charges on electricity bills.

For Boston Edison, this sale means the company is absolved of future responsibility for the plant, including the hassle of the eventual shutdown while consumers pay the debts from past construction and repairs. Of the $1 billion the company's customers are paying for all of Boston Edison's "stranded costs," over $500 million is from the Pilgrim plant alone. Consumers will cover this debt through "transition" charges on their electricity bills.

Like Oyster Creek, the Pilgrim plant wasn't supposed to be a prime candidate for acquisition; it was just too expensive to operate. Deregulation was expected to reveal that reality, not camouflage it. In fact, one of the more centrist environmental groups, the Conservation Law Foundation (CLF), helped successfully fight a referendum last year that would have overturned Massachusetts's deregulation law because the group believed the law would be good for the environment. CLF's campaign literature maintained that competition would "hasten the retirement of the region's old, dirty, and inefficient coal- and oil-fired power plants and its nuclear units." Under this theory, the polluters would then have been replaced with cleaner, combined-cycle gas turbine plants.

Of course, that hasn't played out. Once its capital costs are written off and charged to ratepayers, an already constructed plant has an obvious advantage—you don't have to pay to build it. In the early 1980s, the story was quite different; utilities built lots of power plants in anticipation of high demand that never materialized, explained David Penn, the deputy executive director of the American Public Power Association. Since then, no companies or utilities have built new base-load generating plants, focusing instead on smaller ones that can be operated during times of peak need. So deregulation won't drive environmentally questionable facilities out of business. Rather, many energy analysts and environmentalists predict older plants will continue to operate while more plants are simply added to the mix.

Periodical Bibliography

The following articles have been selected to supplement the diverse views presented in this chapter. Addresses are provided for periodicals not indexed in the *Readers' Guide to Periodical Literature*, the *Alternative Press Index*, the *Social Sciences Index*, or the *Index to Legal Periodicals and Books*.

Rod Adams — "Alternative Nuclear Power," *World & I*, April 2001.

Rosalie Bertell — "Victims of the Nuclear Age," *Ecologist*, November 1999.

Donald A. Blackburn — "An Environmentalist's Choice," *San Francisco Chronicle*, March 25, 2001.

Harold Denton — "Nightmare at Three Mile Island," *George*, March 1999.

Masha Gessen — "The Nuclear Wasteland," *U.S. News & World Report*, February 26, 2001.

Alok Kumar — "Nuclear Power: Blessing or Blight?" *World & I*, May 1997.

Marianne Lavelle — "When the World Stopped," *U.S. News & World Report*, March 29, 1999.

Douglas McGregor — "Rethinking Nuclear Power," *New American*, April 23, 2001.

Sam Nunn — "Managing a Nuclear Transition," *Washington Quarterly*, Winter 2000.

Jonathan Rabinovitz — "A Push for a New Standard in Running Nuclear Plants," *New York Times*, April 11, 1998.

Rebecca Smith — "Two Utility Executives See Potential Riches in Nuclear Stepchildren," *Wall Street Journal*, October 28, 1999.

Michael Steinberg — "Stop McNukes," *Z Magazine*, February 2000.

William Tucker — "More Nukes, Please," *Weekly Standard*, April 2, 2001.

Woody West — "The Waning of Nuclear Power," *Insight*, May 17, 1999.

Bertram Wolfe — "Nuclear Power Is the Answer to Energy Scarcity," *Los Angeles Times*, September 18, 2000.

CHAPTER 3

What Alternative Energy Sources Should Be Pursued?

Chapter Preface

Scientists have developed ways to turn ordinary garbage into energy. "Biomass"—organic refuse such as corn stalks that can be turned into energy—is being used to produce ethanol, which is mixed with gasoline to help power automobiles.

Biomass fuel is a renewable energy source and there-fore—unlike fossil fuels—is inexhaustible. An additional advantage of developing renewable energy sources, which also include solar and wind power, is that the United States could decrease its dependence on foreign oil suppliers. Another benefit of renewable energy sources is that many of these alternatives are less harmful to the environment than are fossil fuels. For example, solar panels transform energy from the sun into electricity without emitting any pollutants. Greenpeace, an organization devoted to exposing and solving global environmental problems, argues that "renewable energy will become an environmental imperative" because the process of using alternatives to generate power does not produce the greenhouse gases that many scientists claim cause global warming.

However, many critics of renewable energy sources argue that such alternatives produce energy far less efficiently than do fossil fuels. For example, energy from the sun is diffuse, which means that solar panels transform only a fraction of the sun's energy into usable power. Indeed, the financial investment in renewable energy sources is often greater than the return, making renewables generally more expensive than energy from fossil fuels. Finally, experts maintain that many renewable energy sources have negative environmental impacts. For example, the turbines that generate power from the wind pose a threat to many endangered birds such as the condor. Jonathan H. Adler, a senior fellow at the Competitive Enterprise Institute, explains that "the problem for wind power is that the same currents that power wind turbines help keep condors, eagles, and other soaring species aloft. Thus the best sites for wind power generation are also the most likely to present bird problems." To be sure, wind power is not the only renewable energy source that can present environmental problems. For example, critics point out that using ethanol to run cars contributes to global warming because all internal combustion engines, regardless of the fuel, produce greenhouse gases like carbon dioxide.

Similarly, using garbage to produce methane for fuel is an ingenious way to produce energy, but while it solves some environmental problems—such as the accumulation of refuse—it can also exacerbate other problems such as global warming, since one end product is still carbon monoxide. Authors in the following chapter debate what alternative energy sources should be pursued. Most experts agree that the best alternatives would be abundant, produce great amounts of power, and have little negative impact on the environment.

> *"Wind farms . . . have little ecological impact. . . . They pose no health risks . . . [and] they will usually pay back the energy cost of manufacture within the first year of operation."*

Why I Hate Wind Farms and Think There Should Be More of Them

Peter Harper

Peter Harper lives and works at the Center for Alternative Technology in Wales. In the following viewpoint Harper contends that choosing wind power is an ethical choice because wind energy does not harm the environment or threaten human health like the burning of fossil fuels does. Furthermore, he claims that the only cost associated with wind power—the marring of the landscape with large turbines—is not imposed upon future generations and other species but is paid by today's energy consumers.

As you read, consider the following questions:

1. Who pays the costs that arise from the combustion of fossil fuels, according to Harper?
2. According to the author, what are two avenues for reducing the consumption of fossil fuels?
3. What are the three environmental values that Harper discusses?

I have been following the wind farms debate ever since it started. In fact long before we had any wind farms in Britain I knew what was coming because I had done the elementary sums. If a modern wind turbine has an average output of one hundred kilowatts and modern conventional power station an output of one million kilowatts, it follows you need 10,000 windmills of this sort to replace one power station. And you cannot pack them all together in a valley somewhere hoping not to notice them: they need to be spread out in the windiest spots, which are nearly always on hilltops in rural areas. So, to make any serious difference to energy supply there have to be an awful lot of windmills, and whatever you do, you can't hide them.

Yes, I knew what was coming, and finally it did. I watched the protest groups and the planning battles and the letters in the newspapers with a sense of unfolding tragedy. But I noticed one thing that I had not anticipated: people who didn't like the visual impact of the windmills also found the noise disturbing, and quite genuinely found the wider environmental arguments for wind power unconvincing; while people who favoured them on environmental grounds somehow didn't mind their appearance, or even found them attractive—and didn't seem to notice any noise at all. This clumping of opinions goes even further: that anti-nuclear activists tend to favour wind farms, while anti-wind activists tend to favour nuclear energy. I rarely meet anybody who does not fall into one of these two camps.

Personally, I seem to be an exception. I really dislike the appearance of wind farms. I live in a hilly part of the country where the scenery is particularly varied and beautiful, and to my eye the ancient skylines are utterly spoiled by this kind of industrial machinery. The slowly-changing tranquillity of distant views is chafed away by fidgeting blades. The illusion of landscape seen through a lens of archetypal vision is quite shattered. Close-up, the noise is not loud, but unnatural and often disturbing. And they need more pylons and transmission-lines to get the electricity distributed. What more can one say? They are just awful!

I also know that the energy generated by a wind farm can be "saved" more cheaply and with far less environmental impact by the most elementary measures of energy efficiency. Anyway the quantity is never very large by conventional power-station standards, so it's worth asking why we should tolerate such dramatic intrusions on our scenery for so little in return.

Yet—listen to this—I support wind power on ethical grounds, and I hope to persuade you to do so as well. I hope also to present a new angle on the wind farm debate, one long overdue.

Let me start at the beginning. Nobody in a modern society can manage without "extra" energy over and above what we can provide from our own bodies, fuelled by toast and potatoes. Even if we don't appear to use very much energy personally, plenty is used on our behalf to provide the hundreds of products and services that flow through a household every year. Even real gung-ho greenies find it hard to live on much less than the equivalent of ten people-power, and most of us use far more.

This "extra" energy has to come from somewhere, and in Britain it comes from fossil fuels: coal, oil and gas, with a bit of nuclear electricity and a dash of hydro from Scotland. Collecting, converting and using this energy has various negative environmental effects, which constitute a kind of "cost" in addition to the financial costs. The important thing about this environmental cost is: who pays it?

In Britain more than 90% of our energy currently comes from fossil fuels. As we all now know, burning fossil fuels has an ineluctable downside: increasing the amount of carbon dioxide in the atmosphere. In essence, every time you use fossil energy, directly or indirectly, you are dumping your waste CO_2 into the global atmosphere. You are collecting the benefits here and now, but slushing off the costs onto everyone else. "Everyone" includes other humans of course, but also many other species, and indeed other generations that will have to cope in the future with the consequences of our present activities. These "costs" may be relatively light, or they may be catastrophic; they are so difficult to assess that for the time being, using fossil fuels is almost the perfect crime: everyone else is in on it too, and you can get away with pretty well anything.

But—there is no getting round it—you are dicing with posterity. The stakes are high.

So using fossil fuels is not a good idea if it can possibly be avoided. We should try to use as little as possible, and certainly much less than at present. There are two avenues for this reduction. The first is to use energy more efficiently, so that less is needed to perform its essential tasks. The second is substituting other sources of energy that do not produce carbon dioxide. As it happens, "using less" while delivering the same level of service is usually cheaper, quicker and easier than using alternative sources, so this should be the cornerstone of our efforts both private and public. But however much we succeed in reducing our demand for "extra" energy, we will always need some. We can plausibly imagine a future Britain in which the total energy "pie" was, say, one-third of its present size, and of this, 60% or so would be met by renewable sources such as solar, wave, tidal, biomass, geothermal, hydro and wind. Although this is very far from current government policy, it is technically quite feasible.

Such a scenario reduces fossil fuel use to between 10% and 20% of its current level in the UK. This would give us a sustainable and globally "fair" quota for CO_2 dumping, and would conserve the dwindling fossil fuels for what they do best: providing instantly-available energy to fill the gaps between the erratic renewables, and guarantee a reliable and equitable energy supply.

Unfortunately this attractive prospect is marred by a growing realisation that the renewables, too, have their environmental impacts.

It is traditional for "environmentalists" to deplore any kind of environmental insult. But the wind farm debates have shown us that there are many kinds of environmentalists, and many different environmental values; and—to our dismay—the different values often conflict. If we look at various sources of energy we see that their environmental impacts differ a great deal, not just in degree but in kind; and our preference for this or that source of energy will depend very much on our deeper tastes and values.

Apart from their effects on the global climate, which could influence just about everything, the fossil fuels tend to affect principally the *natural* environment. Coal is particularly bad (acid rain, open-cast mining), oil is not far behind (tanker spills, more acid rain), while gas is relatively clean, although there is not so much of it to be had as the other fossil fuels. In contrast, nuclear energy has a relatively small effect on the ecology of *natural* systems; its environmental impact is nearly all to do with *human health*. The Welsh sheep that cannot be sold owing to excessive levels of radioactive caesium from Chernobyl are not that bothered by it, neither is the radioactive grass they feed on. But *we* are bothered. Real people are at risk, get cancer; and in the case of a serious accident—unlikely as it is—whole regions may become uninhabitable to humans.

Wind Power

After changing little from the late 1980s through the late 1990s, the nation's large wind farms have markedly increased their electricity production capability since 1998.

Capacity of U.S. wind energy farms (in megawatts)

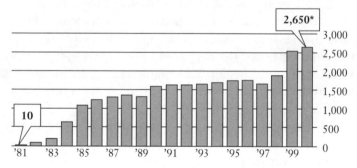

*Projected

Note: A megawatt is enough energy to supply 1,000 homes.

U.S. Department of Energy Wind Energy Program, American Wind Energy Assn.

The contrast here is between nature-centred and people-centred values. They are both worthy of respect; but they do not necessarily coincide.

The "renewable" sources of energy sometimes share these negative consequences: biomass monocultures or tidal barrages can have profound ecological effects; large dam projects can forcibly displace millions of people and bring about new patterns of disease, as well as destroying unique habitats. But the principal impact of most renewables is *visual and aesthetic*. This is simply a consequence of their (usually) low energy-density. Although the total amount of energy potentially available may be very large, it is spread through a huge space. It is so "dilute" that very big, or very many, installations are needed to collect enough to make any difference. This is particularly true of sun and wind, and wind has the added curse of needing to be in exposed upland regions, often in the "unspoiled" west of Britain. This is why the wretched things are so irredeemably visible.

Now which of these environmental values—ecological, health-related, or aesthetic—is to be preferred? I would argue broadly that, other things being equal, if we are faced with a choice of sacrificing nature-values, human-health-values or aesthetic values, the presumption must be against aesthetics, simply because they are to a large extent arbitrary, the result of labile cultural processes. That is not to say they should be ignored, or that they have no weight, but they have (self-evidently?) less weight than risk to life and limb or irreversible loss of habitat and species, or climate change.

That is one ethical guideline. Another is this: if ever you have a chance to pay environmental costs up-front, you should take it.

Surely it is honourable to try and pay the environmental costs as directly as possible, and not to try and shuffle them off onto other people, other species, or other generations? In the case of fossil and nuclear fuels it is almost impossible to pay the full costs. In using fossil or nuclear fuels you are unavoidably imposing costs and risks on other "agents" who have not been asked, and if they were would almost certainly say No.

The renewables avoid many of these ethical difficulties. Wind farms, I believe, avoid them all. They have little ecological impact, most of it transient, at the construction phase. They pose no health risks, apart from the occupational ones in erecting and maintaining the structures. They will usually pay back the energy cost of manufacture within the first year of operation. Their environmental impact—

unlike that of conventional sources—is completely reversible, leaving all options open to our descendants. They can be decommissioned easily, safely, and in a matter of weeks. The components can readily be re-used or recycled. They combine well with traditional land uses such as agriculture. Their environmental impact is almost entirely visual and aesthetic; but—crucially—this "cost" is paid on the nail by everyone who sees them. At first sight this seems like a strike against wind power, and it probably explains the vigour of the objections. But I am arguing that, ethically, this is actually a *positive* quality. Generally it is a feature of the renewable energy sources that their costs and benefits are *congruent:* they accrue to more or less the same populations— certainly on a national scale; while for fossil fuels the costs and benefits fall on *different* populations.

The World's Fastest Growing Energy Source

[Wind] turbines are now so efficient that, according to US government estimates, North Dakota, Texas and Kansas alone could produce far more energy than is needed to power the entire United States (although in reality wind power development would be more widely dispersed across the nation).

Converting to wind power could take a bite out of the US trade deficit, about $80 billion—or 20%—of which pays for oil-related imports. And it would no doubt be welcome among America's beleaguered farmers, who could earn money from the sky above their fields. The US has incentives to promote just that, but analysts say they are too weak. Though wind is the world's fastest-growing source of energy, America has added capacity for fewer than 20,000 households during 2000.

David Case, *Progressive Populist*, March 15, 2001.

This lack of ethical congruence for fossil fuel use is uncomfortable, and generates hypocrisy. Nearly all objectors to wind farms live in rural areas and don't want their landscapes spoiled. This is quite understandable; neither do I. But such people invariably wish to retain the benefits of abundant cheap energy, brought to them discreetly by means of cables, pipelines and tankers, for whom others (mostly in urban areas or overseas) have borne the impact of mining, refining, conversion, transmission etc, not to mention the wider implications of CO_2 release. This is simply hypocritical. In opposing wind power on grounds of personal aesthetics, objectors are implicitly supporting something far worse, and trying to avoid paying their dues.

Which brings me back to where I started. I too live in a rural area. I benefit from cheap and abundant energy. In fact when I actually see a wind farm I am nearly always driving a car, cheerfully enjoying the benefits of fossil fuels, and shamefully slushing the environmental costs off onto the world at large, while trying to think of all sorts of ways to justify what is in the end, indefensible.

Now much as I hate the sight of these monsters in the countryside, they bring me some relief from my appalling hypocrisy. As I gaze upon them, it dawns on me that *I am now paying the environmental bill.* Ethically the only way I could argue against wind power would be to use so little energy myself that I didn't need to buy any, ever. But that is not going to happen, to me or anyone else. So of all the sources I could choose, wind is one of the honest ones, where I can pick up the tab myself, right now. To oppose it on grounds of personal taste would be an act of the grossest and most shameful NIMBYism [an acronym that stands for Not in My Backyard].

So I call on all the well-heeled middle-classes in the countryside to PAY YOUR DUES. The more you loathe the visual impact of windmills, the better, deep down, you should feel when you see them. As for me, whatever I think about the aesthetics of wind power (ugh!) I feel I have no option but to press for more windmills wherever suitable sites can be found. They are the least of many evils.

> *"[Solar] energy is extremely simple and clean, involves no moving parts, needs no support facilities, and has no toxic by-products to conflict with the environment."*

Solar Power Is Becoming Commercially Viable

Barbara Wolcott

In the following viewpoint Barbara Wolcott reports that solar power is becoming commercially viable due to advances in the design of photovoltaic cells, which transform sunlight into electricity. She explains that many analysts support the use of solar energy because it does not harm the environment the way burning fossil fuels does. In addition, experiments with solar power in Sacramento and New York illustrate that municipalities can utilize this emerging energy source to save money. Barbara Wolcott is a freelance writer.

As you read, consider the following questions:
1. According to BP Amoco, how much has the cost of solar cells decreased from 1980 prices?
2. Until recently, what has been the cost of a solar power installation for an individual home, according to the author?
3. According to Wolcott, what engineering problem did the New York Power Authority run into at its Westchester County Wastewater Treatment Plant?

S olar power may finally be shedding its image problem. One of its biggest boosters right now is NASA. Solar cell technology has an important role in the National Aeronautics and Space Administration's Performance Plan for 1999 and beyond. The agency pledges a "faster, better, cheaper" approach to reduce spacecraft development time and costs.

Solar Energy Goes Commercial

But solar power is also getting more backers in the commercial market here on the ground, where banks of cells and even some roofing materials provide primary power for households and contribute to the available electricity supply. A California utility, for instance, has embarked on a program that could eventually supplement its power generation capacity by about 10 megawatts (MW) of photovoltaic energy created right on the roofs of customers' homes. In New York, a program of a different sort has, as one of its long-range aims, a curb on taxes.

Advocates argue that photovoltaic energy is extremely simple and clean, involves no moving parts, needs no support facilities, and has no toxic by-products to conflict with the environment. Basically, you take a substance and expose it to sunlight. The only environmental impact is in the manufacture, which is minimal.

Solar power has its share of marketing problems. Not only does the cost of installation make it more expensive than electricity from traditional sources, but the panels are troublesome to set and, in some cases, ugly. New marketing deals and new product designs are attempting to address those problems.

The solar power movement is only a few years old in the United States, but given its aggressive marketing for the last 15 years in other countries, it is much stronger in Europe and Japan. And it is gathering support in the States.

Signs of strength in the solar technology market have encouraged BP Amoco, for example, to commit significant production volume to the cells. The company operates BP Solar, which has manufacturing plants in six countries, including one in Fairfield, Calif. BP says the cost of solar cells is now one-seventh of what they were in 1980, because of improved manufacturing methods and increased market demand. BP Solar expects its sales will increase to $1 billion in the next decade. BP Solar makes crystalline silicon and thin film modules, and calls itself an integrated solar power supplier; that is, it designs, makes, markets, installs, and services solar power systems.

Sacramento's Solar Experiment

In northern California, the Sacramento Municipal Utility District has become an activist organization in the spread of solar power, chiefly because of the explosive population growth in its area. With more than 500,000 customers, it is the fifth-largest municipally owned electric utility in the United States in terms of customers served. The company, which goes by the nickname of SMUD, has had to address both distribution and production issues that do not face utilities serving longer-established populations, which already have poles and lines widely installed.

Power generation costs about half as much as putting in transmission lines. Running new residential lines costs in the neighborhood of $1 million a mile, and running power to an industrial site can cost significantly more. So SMUD looked seriously at alternative power to reduce capital costs. It polled customers to ask if they were interested in green energy. The answer was emphatically yes and, in 1993, the utility embarked on a solar power project called the PV Pioneer Program.

SMUD offers to install solar panels on the roofs of customers who volunteer to participate. Under the original plan, the utility owns and maintains the solar systems. Homeowners pay the same as their average electric bill plus a surcharge of about $4 a month. Four hundred fifty homes and 20 other sites participate in the program, which contributes 1.5 MW to the power grid.

A new version of the program, PV Pioneer II, offers to sell systems using solar panels or an alternative, photovoltaic roofing, to customers, who own the power that is generated. More than 200 homeowners have applied for the deal, said Sherri Eklof, program manager for the Pioneer II program. She said SMUD has a goal of installing a maximum of 100 systems this year.

Until recently, the cost of a solar power installation for an individual home has been about $18,000 in the open market, but SMUD's customers can purchase a complete 2 kilowatt (kW) solar power system for $4,740. SMUD picks up just over half the actual cost of the installation, so the total price for the job comes in around $10,000. The price, which is for a system using solar panels, includes the arrangements for net metering, which is necessary to sell extra power back to the utility. The utility makes available a 10-year credit plan that charges 9.5 percent.

According to Eklof, three installations have been completed so far under Pioneer II, which was launched last fall.

The solar panels from SMUD's principal supplier, Energy Photovoltaics Inc. of Lawrenceville, N.J., do not use conventional crystal wafers, but instead are made with a thin film coating on glass, a design that reduces the cost of the product.

A secondary supplier to the Pioneer II program is Solarex, a BP Amoco subsidiary based in Frederick, Md.

Photovoltaic Systems

Energy Photovoltaics, or EPV, has a deal to deliver 10 MW of solar power products to SMUD over the five years, through 2002. The utility has a similar arrangement for inverters with Trace Engineering of Arlington, Wash., Eklof said. EPV's product uses amorphous silicon in what it calls photovoltaic modules. Conventional solar panels use crystalline silicon wafers.

According to Alan Delahoy, vice president of research, Energy Photovoltaics creates modules by coating glass with a thin film, about 400 nanometers (nm) thick, or hundreds of times thinner than the usual crystalline wafer, which is about 300 microns thick. The company can coat as many as 48 pieces of glass at a time, Delahoy said. He estimated that advantages in the design and the process cut the cost of manufacture in half.

Most of the product shipped to SMUD so far has come from EPV's factory in Lawrenceville. The company has a new, larger factory in Budapest, where it operates as Duna-Solar. There are plans for another site in SMUD's hometown of Sacramento, where the operating name will be CalSolar.

Delahoy said EPV's lab is working on another thin film, using copper, indium, gallium, and selenium. The National Renewable Energy Laboratory (NREL) in Golden, Colo., set a record last December for electricity conversion with a photovoltaic cell combining those elements. The NREL's thin film cell converted sunlight into electricity with 18.8 percent efficiency.

Dual-Purpose Shingles

The photovoltaic roofing consists of dual-purpose shingles: They keep the rain out and also contain a material that generates electricity from sunlight. Marketed under the brand name Sunslates, they generate 10 watts (W) per square foot and, under SMUD's program, a portion of the roof is covered with enough of them to generate between 2 and 4 kW. The rest of the roof is weatherproofed conventionally.

Eklof said SMUD has installed Sunslates on the roofs of four model homes so far under a pilot program. She said a few retools of occupied homes are in development at the 1998 price of $2.23 a watt for the active portion of the roof. The photovoltaic roofing, supplied by Atlantis Energy Inc. of Colfax, Calif., replaces panels with sleek tiles, not unlike slate or asphalt.

In the roofing tile, six photovoltaic cells are connected with fine metal ribbons and attached to a concrete and fiber backplate that has been coated with conductive metal. The result is a roof tile that is approximately the same thickness as conventional asphalt roofing. The tempered glass tiles have been documented to withstand temperatures ranging from minus 23 [degrees] to 166 [degrees] F, and winds in excess of 125 mph.

A similar product is in development at EPV, which at present makes solar power modules. According to the company's director of international marketing, Eva Csige, EPV has retained an outside firm that is working on a design to apply the company's technology to a photovoltaic shingle.

SMUD has a history of using alternative energy. It maintains several solar power plants, the largest of which is a 10-year-old plant with four units turning out 2.7 MW at Rancho Seco, near a retired nuclear facility.

Creating Goodwill

The utility also produces solar energy in banks of panels on covers that it builds over parking lots. In a region that commonly has summer temperatures in triple digits, the goodwill created by the shade from those roofs is in itself a public relations coup for SMUD.

Sacramento's growing number of small rooftop power plants help the utility meet its electricity demand at the peak hours between 11 A.M. and 2 P.M., because they are the same hours when the photovoltaic systems are producing excess power and sending it to the power grid. That additional power has led the utility to put off or avoid building more full-size plants.

© Kirk Anderson. Used with permission.

The New York Power Authority (NYPA), meanwhile, has solar power installations at 11 sites in New York City and Westchester County, and plans seven others, on Long Island, in Buffalo, and northwest of Albany, the state capital. The largest is a 330-kW system at New York City Transit's Gun Hill bus depot in the Bronx.

The Power Authority supplies more than one-quarter of New York State's electricity, directly serving government and business customers and selling it wholesale to municipal systems, rural cooperatives, and investor-owned utilities throughout the state. According to a spokesman, the Power Authority's primary market lies in New York's public sector. Currently, the NYPA isn't making money on its solar installations. Instead, they are intended to demonstrate "that the technology is here." And also to prove that solar power can work in a climate like New York's, which gets perhaps half as much sunlight as the Southwestern states.

At present, NYPA puts the cost of installing solar power equipment at about $8,000 per kilowatt. The authority estimates that, with 20-year financing and the agency's bond interest rate, solar power costs NYPA about twice as much as electricity generated by more conventional means. In other words, solar power in New York at current prices carries a 40-year payback.

NYPA expects costs to drop as technology develops and the volume of installations increases. When that happens, power bills at schools, libraries, and other public buildings, and the tax money needed to pay them, may be reduced by the addition of solar power.

Each job goes out to bid separately, said Lumas Kendrick, the research engineer for the solar project. NYPA has no long-term primary supplier as SMUD does. But, according to officials at the agency, that is part of the learning process, which is the value of the program so far.

The Power Authority has said that the experience it receives by installing various systems will prove very valuable when manufacturing and market influences drive prices down, and solar power becomes a profitable enterprise for the agency.

Unexpected Challenges

In the course of its program, the authority has run into some engineering challenges it didn't expect. It lost 30 panels to high winds, for instance, at its second-largest installation, a 110-kW system at the Westchester County Wastewater Treatment Plant in Yonkers. The authority's president, Eugene Zeltmann, said of the incident in a speech in 1998, "Even though the broken panels had to be replaced, we learned a valuable lesson on how to site and install photovoltaic (PV) systems."

A 36-kW system at a composting facility on Rikers Island in New York City was installed in 1996. The authority found that the escaping gases caused equipment to rust faster than at other sites. The NYPA plans to put housings over sensitive parts to shield them from condensation, according to Kendrick.

NYPA's solar energy installations generate a portion of the electricity needs of a building. Zeltmann said that the Gun Hill installation at peak output is able to meet half the electricity needs of the bus depot.

The Gun Hill depot site generates electricity worth about $2,500 a month for the city's transit authority, and the use of that much solar power keeps an estimated 600,000 lbs. of carbon dioxide, 1,200 lbs. of NO_x, and 400 lbs. of particulates out of the air each year.

The authority has placed 6- to 8-kW solar generators at some public schools, and is working with district officials in the state to add more.

| "Someday soon, a washing-machine-sized fuel cell might be quietly at work powering your home."

Fuel Cells Are a Promising Energy Source

William McCall

William McCall reports in the following viewpoint that utilities, automakers, and government agencies are investing in fuel-cell technology that may one day eliminate the country's reliance on fossil fuels. According to McCall, fuel cells run on hydrogen, which is the most abundant element in the universe and burns without harming the environment. Although fuel cells are expensive, innovators hope to improve the technology to make the cells affordable enough to power automobiles, home energy systems, and electronic equipment. William McCall writes for the *Associated Press*.

As you read, consider the following questions:

1. What famous example of NASA's use of fuel cells have most Americans heard of, according to McCall?
2. According to the author, what have ONSI fuel cells traditionally been used for?
3. What is the price of the home-use fuel cell currently being tested by Idacorp and the Bonneville Power Authority, according to McCall?

It's called the hydrogen economy, and we're on the brink of it. Automakers are in a race to replace the internal-combustion engine with fuel cells. And someday soon, a washing-machine-sized fuel cell might be quietly at work powering your home—and perhaps even feeding some electricity back to the grid.

No Longer Science Fiction

As the lights dim in parts of the nation because of an energy shortage, the model for supplying clean and abundant electricity in the 21st century can be found at a Portland, Oregon sewage plant.

Methane collected from decomposing waste provides hydrogen to power a commercial fuel cell that transforms the volatile gas into enough electricity to light more than 100 homes for a year.

Next year, King County [in Washington State] will start a fuel-cell generating operation, five times bigger than Portland's and costing $18.8 million, at its wastewater-treatment plant in Renton.

Fuel cells were invented in the 19th century. But most Americans had never heard of them until a faulty one blew a hole in Apollo 13 in 1970, scuttling what would have been the third moon landing and nearly costing three astronauts their lives.

Fuel cells remain an essential part of the space program, reliably powering the space shuttle.

And now utilities, investors, government planners and automakers are envisioning some down-to-Earth uses for a technology that converts the most abundant element in the universe—hydrogen—into electricity and water.

"It's no longer science fiction," said Steve Millett, one of the leading researchers in the field. "It's real."

Cars Got the Wheels Turning

Millett works at Battelle, the institute founded by a steel-industry family in Columbus, Ohio, which now develops all kinds of technology for industry and the government, including the National Aeronautics and Space Administration (NASA).

Millett says fuel-cell technology was transformed during the past decade from a cottage industry into one of the most rapidly expanding high-tech businesses in the world, partly due to the automotive industry's sudden keen interest.

"More and more auto companies have awakened to the fact that their sales are dependent on fuel prices," Millett said, "so the auto companies are investing more in fuel cells and pushing harder than any other industry."

As 2001 began, Exxon Mobil planned to join Toyota and General Motors in an alliance to develop environmentally friendly fuel-cell vehicles. Ford, Honda, Nissan, BMW, Daimler-Chrysler, you name it—all have fuel-cell projects in the works.

Fuel-cell-powered cars are currently being road-tested on automakers' tracks. Fuel-cell buses have hauled passengers in long-term tests on the streets of Chicago and Vancouver.

And Honda is already promoting its fuel-cell car in a television commercial.

A Fuel Cell in Every Home

Yet as recently as 1996, there was only one major manufacturer of commercial-size fuel cells in the country—ONSI Corp. in Windsor, Conn., a subsidiary of International Fuel Cells.

ONSI built units the size of a minivan to provide electricity to facilities that were too remote from the main power grid or needed reliable backup power, such as hospitals and resorts.

Now dozens of manufacturers and many large companies are considering fuel-cell development in an industry that has one of the fastest-growing trade associations in the country—the U.S. Fuel Cell Council in Washington, D.C.

"There are a lot of big names in the business now," said Bob Rose, the council's director. "General Electric is in, along with 3M, DuPont, United Technologies."

In the 1960s and 1970s, utilities were interested in building big fuel-cell plants capable of producing one to three megawatts as part of a central power supply. But the long-range goals have shifted to a smaller scale: putting a washing-machine-sized fuel cell in every home, or smaller units in every car and truck. And that's attracted a broader range of companies and investors, Rose said.

Motorola's Project

Motorola, for example, is working with the Los Alamos National Laboratory in New Mexico to miniaturize fuel cells for handheld devices like cell phones.

Mark Williams, fuel-cell product manager at the National Energy Technology Laboratory in Morgantown, W.Va., said the U.S. Department of Energy—the lab's parent agency—has been funding fuel-cell research across the country for years. But now initial public offerings of stock in various companies are spurring investment interest, along with research funding by established companies.

Not Just Science Fiction Anymore

Way back in 1874, French writer Jules Verne published a science fiction novel, *The Mysterious Island*, in which a character predicts that a certain fuel will one day "furnish an inexhaustible source of heat and light." The name of that fuel: hydrogen.

More than a century later, Verne's prediction has yet [to come true]. But that could soon change. Hydrogen is poised to take off as the main energy source of this century, powering everything from cell phones to cars.

Hugh Westrup, *Current Science*, April 6, 2001.

"The alliance of auto manufacturers, fuel-cell developers, utilities, universities—there's a whole new initiative that's bringing it together," Williams said.

Cell Percolates in Portland

At the Portland sewage-treatment plant, at the confluence of the Willamette and Columbia rivers, the city is generating electricity from only the third commercial fuel cell of its kind in the nation to use waste "biogas."

The fuel cell began operating in July 1999. It's such a success that the Environmental Protection Agency gave the city a "clean-air excellence" award for converting waste gas from sewage into 200 kilowatts.

David Tooze, energy program manager for Portland, said the city needed to pull together several grants to cover the $1.3 million cost of the fuel cell. But it has proved to be worth the investment by producing electricity at 8 cents per kilowatt hour, when the deregulated spot market easily pushes the price to 20 cents per kilowatt hour.

Fuel cells are an extremely clean power source because they combine hydrogen (derived from sources such as methanol, petroleum products, natural gas or renewable resources) and oxygen—the two elements that make up water, the main byproduct.

"A lot of long-term energy planners recognize fuel cells could be one of the major links that bridge us from a society that operates on fossil fuels and their pollution liability, to an energy economy that operates on hydrogen, which is essentially clean-burning," Tooze said.

Northwest Is a Leader

Much of the work is taking place in the Pacific Northwest.

Ballard Power Systems, a pioneering company, is based in Vancouver, B.C.

And a public power consortium of 13 utilities called Energy Northwest is taking part in a Bonneville Power Administration (BPA) test of fuel cells made by a company founded in the central Oregon town of Bend.

Bonneville officials see fuel-cell technology as a way to redistribute the power grid on a more local level. The federal power marketing agency already oversees one of the cleanest energy sources in the nation, the string of 29 hydroelectric dams on the Columbia and Snake rivers.

But salmon conservation and increasing power demand have forced the BPA to search for alternate sources of electricity. Fuel cells could be a way to take the load off the central power grid—a system of residential generators could power homes and even send surplus to the grid.

Cells Are Costly for Now

But fuel cells are still expensive.

For example, a Boise-based company called Idacorp has joined the BPA to test home-use fuel cells that are in the

$25,000 range. The company hopes the cost per unit eventually will drop to the $5,000 to $7,000 range.

Still, fuel-cell development is key to energy independence.

"It's going to take time, and it's not effortless, but the move toward the so-called hydrogen economy has started," said Millett, the Battelle researcher.

> *"Methane hydrates hidden beneath U.S. waters alone hold some 200 trillion cubic feet of natural gas, enough to supply all the nation's energy needs for more than 2,000 years."*

Methane Hydrates May Provide Energy in the Future

Richard Monastersky

Richard Monastersky reports in the following viewpoint that methane hydrates found beneath the ocean floor and in polar permafrost could meet the world's future energy needs. Methane hydrate deposits—which are formed when bacteria excrete methane, which then becomes trapped in sediment—contain natural gas, which is a clean-burning energy source. Although Japan and other countries are exploring ways to extract the methane deposits, questions remain about whether the hydrates can be extracted safely and economically. Richard Monastersky is a staff writer for *Science News*.

As you read, consider the following questions:

1. According to the author, why is Japan so interested in exploring methane hydrates?
2. What is seismic reflection profiling, according to Monastersky?
3. According to the author, what are some of the dangers associated with the drilling of methane hydrates?

Excerpted from "The Ice That Burns," by Richard Monastersky, *Science News*, November 14, 1998. Copyright © 1998 by Science Service Inc. Reprinted with permission from *Science News*, the weekly newsmagazine of science.

C an methane hydrates fuel the 21st century?
In October 1999, the Japan National Oil Corp. will send a ship to a spot about 60 kilometers off a cape called Omae zaki, not far from Mount Fuji. Its crew will lower a drill through 950 meters of water and then start cutting a circular hole the width of a dinner plate into the seafloor. At first, the bit will slice through fine silt as soft as birthday cake. Then, at a depth not yet known, the diamond-tipped drill will breach a hard icelike layer and, in the process, reach into the postpetroleum future.

From Nuisance to Resource

The frozen substance is called methane hydrate, a name that has been increasingly echoing off the walls of Congress, university research offices, and oil company conference rooms around the world. Found under the ocean floor and polar permafrost, methane hydrates are a crystalline combination of natural gas and water, locked together into a substance that looks remarkably like ice but burns if ignited. Until recently, the natural gas industry considered it only a nuisance, something that occasionally plugs up pipelines. Now, some scientists view methane hydrates as the resource that may power the 21st century, and governments are scrambling to explore its promise.

"Methane hydrates are a potentially enormous natural gas resource," declared a U.S. presidential commission in 1997 in its report on future energy research. "It may be that [natural] gas can be produced economically from the methane hydrates on the continental shelf, and this may prove to be a very large new source globally, particularly for some developing countries such as India as well as for the United States," concludes the report.

With some geologists predicting that oil supplies will tighten in the next 15 years, the prospect of vast new fossil fuel deposits has fired the imagination of energy experts. According to some estimates, the energy locked within methane hydrates amounts to more than twice the global reserves of all conventional gas, oil, and coal deposits combined. The U.S. Geological Survey (USGS) estimates that the methane hydrates hidden beneath U.S. waters alone hold some 200 trillion cubic feet of natural gas, enough to supply all the nation's energy needs for more than 2,000 years at current rates of use.

Lured by such a vast resource, Congress is considering a bill in 1998 that would establish a national methane hydrates research program.[1] At the same time, the U.S. Department of Energy is proposing a plan aimed at making it possible to extract methane commercially from hydrates in less than 20 years. Canada, India, Korea, and Norway have all joined Japan by initiating their own hydrates research programs.

Drill-Pipe Dreams?

Such hopes may be little more than drill-pipe dreams, though. At this point, no company or government has demonstrated how to pull natural gas out of methane hydrates deposits without pouring a tremendous amount of money down the borehole.

"The bottom line is there is a lot of gas hydrate. There's probably more gas hydrate than all other resources. But because we have no sense of how much gas hydrate is actually recoverable, we have to be careful," says Timothy S. Collett, a geologist who studies methane hydrates at the USGS in Denver. "It may be totally irrelevant to any resource issue," he says.

Japan has taken the lead on methane hydrates exploration because its geologic heritage has left it with few options. "We don't have many energy resources near Japan, so we mainly import oil and gas from foreign countries," explains Arata Nakamura, assistant project director at the Japan National Oil Corp. (JNOC), a quasi-governmental company headquartered in Tokyo. "JNOC is very interested in conducting research to develop methane hydrates," he says.

Just how deep that interest runs is a matter of some secrecy. Like other oil and gas companies, JNOC considers many exploration issues proprietary. In 1994, Japan's Ministry of International Trade and Industry established a 5-year methane hydrates research plan, culminating in the offshore-drilling project slated for next year. Nakamura declined to

1. An amended bill, which passed the Senate on April 19, 1999, directed the secretary of energy to coordinate a research and development program to develop methane hydrate resources.

specify exactly how much the plan will cost but said it will total more than $60 million.

In a preview of next year's program, JNOC funded a drilling operation in February and March of 1998 at an inland site in the Mackenzie Delta of northwest Canada. Working with the Geological Survey of Canada, a Japanese team bored a well 1,150 meters deep into the Arctic permafrost, where methane hydrates are common.

Permafrost Treasures

Using a hollow drill, they pulled up cores of sandy sediment that formed the ocean bottom many millennia ago. This once-soft sand was as solid as concrete. In places, methane hydrates filled almost all the space between the sand grains, cementing the sediment into frozen layers located between 900 meters and 1,100 meters below the surface.

The main purpose in drilling the well, called Mallik 2L-38, was to measure how much hydrates hide in the sediments, says Scott Dallimore of the Geological Survey of Canada in Ottawa, who coordinated the scientific research at Mallik. "Methane hydrates occur in very high concentrations in the Mallik well, much higher [concentrations] than have been observed anywhere else," he says.

In sand, explains Dallimore, the grains occupy only about 65 percent of the total space, leaving a network of pores that take up the rest of the volume, much like the gaps between the nuts in a jar of cashews. At Mallik, methane hydrates fill 55 percent of the pore space surrounding the sand grains. So, roughly 20 percent of each coreful of the cemented sand was methane hydrates. That percentage exceeds the concentration found even in the richest known deposits under the seafloor, says Dallimore.

The Japanese team chose to drill onshore at Mallik because it is easier to work on land than under the ocean. When it comes to searching for domestic methane hydrates, however, JNOC must move underwater. Japan has no permafrost of its own.

The lair of most methane hydrates lies far from shore and deep below the waves. There, in water depths of at least 600 to 800 meters, low temperatures and extreme pressures in the sediments combine to squeeze methane and water into a crystalline structure. Each molecule of methane gets trapped within a cagelike lattice of frozen water molecules, an arrangement that greatly concentrates a large amount of methane into a small space. Hydrates also go by the name of clathrates, a term derived from the Latin word for lattice.

The methane in most hydrate deposits originally comes from bacteria living beneath the seafloor. As they consume bits of plant and animal remains in the sediment, the bacteria excrete methane, a process still going on today. "It's the same thing as swamp gas or sewer gas," says Roy D. Hyndman of the Geological Survey of Canada in Sidney, British Columbia. When conditions are cold and the pressure is high, the bacterial gas gets locked up into hydrates.

Burning Ice

A chunk of methane hydrate is dense and milky, yet it floats like ordinary water ice. Released from the ocean floor, methane hydrate rises, fizzing like an Alka Seltzer as it discharges its gaseous contents. Frozen methane hydrate reaching the surface can be ignited, hence the term "the ice that burns."

Steve LaRue, *The San Diego Union-Tribune*, March 28, 2001.

In some deposits, the source of the gas lies much deeper, in sediments warmed by Earth's internal heat. Several kilometers below the sea floor, the temperature in the sediment rises so high that it cooks the buried organic debris. This slow simmer produces petroleum and hydrocarbon gases, which leak upward toward the seafloor. As the gases reach cooler sediments, they can form hydrates containing a mixture of hydrocarbons.

Unconcentrated Methane

Or so the theory goes. Methane hydrates lie so deep beneath Earth's surface that geologists are uncertain about even the most basic details concerning the deposits. "We have to understand where they occur and why they occur," says Collett. The Energy Department, in its plan for a 10-year research program, puts the top priority on resource characterization—in other words, determining how much is out there and how to locate the richest deposits.

The same methane locked up in hydrates also comes out of the rear ends of cows and sheep, but the gas industry is not bursting down barn doors to collect the flatus of farm animals. Each animal produces only a little methane, and companies would go bankrupt trying to collect sizable quantities of gas. Likewise, most methane hydrate deposits are probably uneconomical because the gas is not concentrated in large-enough amounts, says Arthur H. Johnson, a geologist with Chevron USA Production Co. in New Orleans.

"While the published estimates of methane hydrate abundance are enormous, it is likely that most of the hydrate occurs in low concentrations and has no commercial potential. Our goal is to be able to find locations where the methane hydrates are sufficiently concentrated to warrant commercial production," Johnson testified at a hearing of the House Energy and Environment Subcommittee in September 1998.

Reflection Profiling

To spot underwater hydrate deposits, geologists rely principally on a technique routinely used by companies searching for petroleum. Blasts set off near the ocean surface send out sound waves that reflect off deep geologic structures and then return to the ocean surface, where they are recorded.

This process, called seismic reflection profiling, sometimes picks up a distinct band in the sediments that parallels the contour of the seafloor. Geologists call these "bottom simulating reflectors." The band marks the bottom of a hydrate deposit, where bubbles of methane gas in the sediment have become trapped below the impermeable frozen layer.

A related technique spots hydrates by estimating the speed of the sound waves as they penetrate the seafloor. In places where hydrates have stiffened up the otherwise soft sediment, sound travels much faster, says Hyndman. He and his colleagues at the Geological Survey of Canada recently used these techniques to pinpoint the potential locations of hydrate deposits off the coast of India.

People in both India and Japan pay three to four times as much for natural gas as do consumers in the United States.

The steep price provides an incentive for these countries to pursue methane hydrates, whereas countries with domestic sources of hydrocarbons currently find hydrates far too expensive to seek.

In the 1999 drilling, Japan planned primarily to extract cores of methane hydrate, which will help in assessing the richness of the deposit, according to Nakamura. The company says it does not intend to "produce" the hydrate, a term engineers use to mean pulling commercial quantities out of the ground.

Nonetheless, U.S. geologists believe that Japan is taking steps toward that goal. At a meeting in 1997 in England, a JNOC official said that after drilling in 1999, the company hopes to take the commercially significant step of classifying some of the hydrate resource as energy reserves, says Michael D. Max, a geologist with the Naval Research Laboratory in Washington, D.C. "What that means is [the resource] changes from a possibility to a certainty. That means they would be able to put some recovery numbers on it, and they can start looking at the commerciality and the costs," says Max.

Extraction Problems

Right now, the costs of producing methane hydrates remain a big question mark because nobody has tried to extract this resource, with the possible exception of the operators of a controversial well in Siberia. Solid hydrates won't come out of the ground as easily as oil and so-called conventional gas, which can flow through rock pores and then up through the drill pipe.

One way to pry hydrates loose would be to release some pressure on the deposit, which would cause the methane and water to split apart, or dissociate. The advantage of this technique is that it would be relatively cheap, says Collett. To relieve pressure, a drill crew could tap the methane gas that often accumulates underneath and pushes up on the deposit. Unfortunately, this process might work too slowly, he says. As hydrates dissociate, they cool down, which stabilizes them and prevents more hydrate from melting.

To speed up the process, crews could drill far below the methane hydrates and pump hot water upward into the deposit, thereby melting the hydrates. Or, they could inject antifreeze from the surface to spur dissociation. "But when you look at the total balance sheet of the issue," says Collett, "the minute you start looking at enhanced techniques, you're putting energy and money into the project, and gas is not a real expensive commodity. So, you end up with the problem that you're putting more money in than you're going to get out in the form of gas."

Even though hydrates remain uneconomical at present, U.S. policy makers see other reasons for researching these deposits. Oil-drilling operations in the Gulf of Mexico are now moving into water more than 1,000 meters deep and are starting to drill through methane hydrate layers more frequently, raising safety concerns. A drill spinning through the hydrate can cause it to dissociate, and each liter of melted hydrate releases 160 liters of gas, says Robert S. Kripowicz, acting assistant secretary for fossil energy at the U.S. Department of Energy.

The freed gas can explode out of the hole, causing the drilling crews to lose control of the well, a costly problem to solve.

"Offshore operators are increasingly reporting problems of drilling through hydrates," Kripowicz told the House energy subcommittee.

Engineers are exploring whether unstable hydrate layers could give way beneath oil platforms or even play a role in triggering tsunamis. Climate researchers have also grown concerned about hydrates because global warming could melt some shallow methane deposits, releasing millions of tons of this potent greenhouse gas into the air.

With so little known about methane hydrates, energy experts say that it is hard to predict whether society will ever tap into these deposits as a fuel source. Still, the Japanese initiative has spurred other oil companies to take an active interest. At a meeting in 1998 in Chiba City, Japan, a group from Shell International Exploration and Production, B.V., discussed its analysis of exploiting methane hydrates. "Our consensus is there are no show stoppers. There is nothing that we cannot handle technically. If we encountered a good accumulation of natural gas hydrates, we could develop it with the existing technology," says Wim J. A. Swinkels, a member of Shell's gas hydrate team. The only issue standing in the way right now, he says, is economics.

Yet, the days of plentiful oil and gas are numbered, and countries will require new energy sources to keep the wheels of progress spinning. "We're enjoying a wonderful economy right now, largely because of the very low cost of energy," said Rep. Vernon J. Ehlers (R-Mich.) at the recent hearing on methane hydrates. "I'm very worried about what's going to happen when the cheap oil is gone, and we're not paying enough attention to it."

| *"Generating electricity with geothermal energy helps to conserve nonrenewable fossil fuels and reduce emissions that harm our atmosphere."*

Geothermal Energy Is Abundant and Clean

Marilyn L. Nemzer, Anna K. Carter, and Kenneth P. Nemzer

Marilyn L. Nemzer is executive director of the Geothermal Energy Office (GEO), an organization that promotes public understanding about geothermal resources. Anna K. Carter is principle of the Geothermal Support Services in Santa Rosa, California and Kenneth P. Nemzer, former chair of the Geothermal Resources Association in Tiburon, California, is an attorney. In the following viewpoint the authors report that geothermal energy—obtained from heat stored within the earth—can help developed countries cut their reliance on foreign oil and help the economies of developing countries grow. Heat or hot water piped from below the earth's surface where magma is rising can generate electricity or be put to direct use for agriculture, heating, and industrial production.

As you read, consider the following questions:

1. What is a geothermal reservoir, according to the authors?
2. What are the three kinds of geothermal power plants, according to the authors?
3. According to the authors, in what three geologic areas are geothermal reservoirs usually found?

Excerpted from "Geothermal Energy Facts," by the Geothermal Education Office, http://geothermal.marin.org, December 23, 2000. Copyright © 2000 by the Geothermal Education Office. Reprinted with permission.

The word *geothermal* comes from the Greek words *geo* (earth) and *therme* (heat), and means the heat of the Earth. Earth's interior heat originated from its fiery consolidation from dust and gas over 4 billion years ago and is continually regenerated from the decay of radioactive elements that occur in all rocks.

Earth's Heat and Volcanic Regions

It is almost 6,500 kilometers (4,000 miles) from the surface to the center of the Earth, and the deeper you go, the hotter it gets. The outer layer, the crust, is three to 35 miles thick and insulates us from the hot interior.

From the surface down through the crust the normal temperature gradient (the increase of temperature with the increase of depth) in the Earth's crust is 17–30°C per kilometer of depth (50–87°F per mile). Below the crust is the mantle, made of highly viscous, partially molten rock with temperatures between 650 and 1,250°C (1,200–2,280°F). At Earth's core, which consists of a liquid outer core and a solid inner core, temperatures may reach 4,000–7,000°C (7,200–12,600°F).

Since heat always moves from hotter regions to colder regions, the Earth's heat flows from its interior toward the surface. This outward flow of heat from Earth's interior drives convective motion in the mantle rock which in turn drives plate tectonics—the "drift" of Earth's crustal plates that occurs at 1 to 5 cm per year (about the rate our fingernails grow). Where plates move apart, magma rises up into the rift, forming new crust. Where plates collide, one plate is generally forced (subducted) beneath the other. As a subducted plate slides slowly downward into regions of ever-increasing heat, it can reach conditions of pressure, temperature and water content that cause melting, forming magma. Plumes of magma ascend by buoyancy and force themselves up into (intrude) the crust, bringing up vast quantities of heat.

Where magma reaches the surface it can build volcanoes. But most magma stays well below ground, creating huge subterranean regions of hot rock sometimes underlying areas as large as an entire mountain range. Cooling can take from 5,000 to more than 1 million years. These shallow regions of relatively elevated crustal heat have high temperature gradients.

Perhaps the best known of these volcanic regions are in the countries that border the Pacific Ocean—the geologically active area known as the Ring of Fire—where the oceanic plates are being subducted under the continental plates. Other volcanic chains form along mid-ocean or continental rift zones (where plates move apart)—in places such as Iceland and Kenya, or over hot spots (magma plumes continuously ascending from deep in the mantle) such as the Hawaiian Islands and Yellowstone.

Formation of Geothermal Reservoirs

In some regions with high temperature gradients, there are deep subterranean faults and cracks that allow rainwater and snowmelt to seep underground—sometimes for miles. There the water is heated by the hot rock and circulates back up to the surface, to appear as hot springs, mud pots, geysers, or fumaroles.

If the ascending hot water meets an impermeable rock layer, however, the water is trapped underground where it fills the pores and cracks comprising 2 to 5% of the volume of the surrounding rock, forming a geothermal reservoir. Much hotter than surface hot springs, geothermal reservoirs can reach temperatures of more than 350°C (700°F), and are powerful sources of energy.

Accessing Geothermal Energy

If geothermal reservoirs are close enough to the surface, we can reach them by drilling wells, sometimes over two miles deep. Scientists and engineers use geological, electrical, magnetic, geochemical and seismic surveys to help locate the reservoirs. Then, after an exploration well confirms a reservoir discovery, production wells are drilled. Hot water and steam shoot up the wells naturally (or are pumped to the surface) where—at temperatures between around 120–370°C (250–700°F)—they are used to generate electricity in geothermal power plants. Shallower reservoirs of lower temperature—21–149°C (70–300°F)—are used directly in health spas, greenhouses, fish farms, and industry and in space heating systems for homes, schools and offices.

Generating Electricity: Geothermal Power Plants

In geothermal power plants, we use the natural hot water and steam from the Earth to turn turbine generators to produce electricity. Unlike fossil fuel power plants, no fuel is burned. Geothermal power plants give off water vapor, but have no smoky emissions.

Flashed Steam Plants. Most geothermal power plants operating today are "flashed steam" power plants. Hot water from production wells is passed through one or two separators where, released from the pressure of the deep reservoir, part of it flashes (explosively boils) to steam. The force of the steam is used to spin the turbine generator. To conserve the water and maintain reservoir pressure, the geothermal water and condensed steam are directed down an injection well back into the periphery of the reservoir, to be reheated and recycled.

Beneficial Impacts

Electricity produced from geothermal resources in the U.S. prevents the emission of 22 million tons of carbon dioxide, 200,000 tons of sulfur dioxide, 80,000 tons of nitrogen oxides, and 110,000 tons of particulate matter every year compared to conventional coal-fired plants.

U.S. Department of Energy,
www.eren.doe.gov/geothermal/geoimpacts.html, April 2001.

Dry Steam Plants. A few geothermal reservoirs produce mostly steam and very little water. Here, the steam shoots directly through a rock-catcher and into the turbine. The first geothermal power plant was a dry steam plant, built at Larderello in Tuscany, Italy in 1904. The power plants at the Larderello dry steam field were destroyed during World War II, but have since been rebuilt and expanded. That field is still producing electricity today. The Geysers dry steam reservoir in northern California has been producing electricity since 1960. It is the largest known dry steam field in the world and, after 40 years, still produces enough electricity to supply a city the size of San Francisco.

Binary Power Plants. In a binary power plant, the geothermal water is passed through one side of a *heat exchanger,* where it's heat is transferred to a second (binary) liquid, called a working fluid, in an adjacent separate pipe loop. The working fluid boils to vapor which, like steam, powers the turbine generator. It is then condensed back to a liquid and used over and over again. The geothermal water passes only through the heat exchanger and is immediately recycled back into the reservoir.

Although binary power plants are generally more expensive to build than steam-driven plants, they have several advantages: 1) The working fluid (usually isobutane or isopentane) boils and flashes to a vapor at a lower temperature than does water, so we can generate electricity from reservoirs with lower temperatures. This increases the number of geothermal reservoirs in the world with electricity-generating potential. 2) The binary system uses the reservoir water more efficiently. Since the hot water travels through an entirely closed system it results in less heat loss and almost no water loss. 3) Binary power plants have virtually no emissions.

Hybrid Power Plants. In some power plants, flash and binary processes are combined. An example of such a hybrid system is in Hawaii, where a hybrid plant provides about 25% of the electricity used on the Big island.

Geothermal Power Production Worldwide

As of 1999, 8,217 megawatts of electricity were being produced from some 250 geothermal power plants running day and night in 22 countries around the world. These plants provide reliable base load power for well over 60 million people, mostly in developing countries.

About 2,850 megawatts of geothermal generation capacity is available from power plants in the western United States. Geothermal energy generates about 2% of the electricity in Utah, 6% of the electricity in California and almost 10% of the electricity in northern Nevada. The electrical energy generated in the United States from geothermal resources is more than twice that from solar and wind combined.

Direct (Non-Electrical) Uses of Geothermal Water

Shallower reservoirs of lower temperature—21–149°C (70–300°F)—are used directly in health spas, greenhouses, fish farms, and industry and in space heating systems for homes, schools and offices.

It is only during the last century that we have used geothermal energy to produce electricity. But using geothermal water to make our lives more comfortable is not new: people have used it since the dawn of mankind. Wherever geothermal water is available, people find creative ways to use its heat.

Hot Springs Bathing and Spas (Balneology)

For centuries, peoples of China, Iceland, Japan, New Zealand, North America and other areas have used hot springs for cooking and bathing. The Romans used geothermal water to treat eye and skin disease and, at Pompeii, to heat buildings. Medieval wars were even fought over lands with hot springs. Today, as long ago, people still bathe in geothermal waters.

In Europe, natural hot springs have been very popular health attractions. The first known "health spa" was established in 1326 in Belgium. (One resort was named "Espa" which means "fountain." The English word "spa" came from this name.) All over Eurasia today, health spas are still very popular. Russia, for example, has 3,500 spas.

Japan is considered the world's leader in balneology. The Japanese tradition of social bathing dates back to ancient Buddhist rituals. Beppu, Japan, has 4,000 hot springs and bathing facilities that attract 12 million tourists a year. Other countries with major spas and hot springs include New Zealand, Mexico and the United States.

Agriculture

Geothermal resources are used worldwide to boost agricultural production. Water from geothermal reservoirs is used to warm greenhouses to help grow flowers, vegetables and other crops. For hundreds of years, Tuscany in Central Italy has produced vegetables in the winter from fields heated by natural steam. In Hungary, thermal waters provide 80% of the energy demand of vegetable farmers, making Hungary the world's geothermal greenhouse leader. Dozens of geothermal greenhouses can also be found in Iceland and in the western United States.

Aquaculture

Geothermal aquaculture, the "farming" of water-dwelling creatures, uses natural warm water to speed the growth of fish, shellfish, reptiles and amphibians. This kind of direct use is increasing in popularity. In China, for example, geothermal aquaculture is growing so fast that fish farms cover almost 2 million square meters (500 acres). In Japan, aqua farms grow eels and alligators. In the U.S. aquafarmers in Idaho, Utah, Oregon and California grow catfish, trout, alligators, and tilapia—as well as tropical fish for pet shops. And Icelanders hope to raise as many as two and a half million abalone a year.

Industry

The heat from geothermal water is used worldwide for industrial purposes. Some of these uses include drying fish, fruits, vegetables and timber products, washing wool, dying cloth, manufacturing paper and pasteurizing milk. Geothermally heated water can be piped under sidewalks and roads to keep them from icing over in freezing weather. Thermal waters are also used to help extract gold and silver from ore and even for refrigeration and ice-making.

Heating/District Heating

The oldest and most common use of geothermal water, apart from hot spring bathing, is to heat individual buildings, and sometimes entire commercial and residential districts.

A geothermal district heating system supplies heat by pumping geothermal water—usually 60°C (140°F) or hotter—from one or more wells drilled into a geothermal reservoir. The geothermal water is passed through a heat exchanger which transfers the heat to water in separate pipes that is pumped to the buildings. After passing through the heat exchanger, the geothermal water is injected back into the reservoir where it can reheat and be used again.

In the Paris basin in France, historic records show that geothermal water from shallow wells was used to heat buildings over six centuries ago. An increasing number of residential districts there are being heated with geothermal water as drilling of new wells progresses.

The first district heating system in the United States dates back to 1893, and still serves part of Boise, Idaho. In the western United States there are over two hundred and seventy communities that are close enough to geothermal reservoirs for potential implementation of geothermal district heating. Eighteen such systems are already in use in the United States—the most extensive in Boise, Idaho and San Bernardino, California.

Because it is a clean, economical method of heating buildings, geothermal district heating is becoming more popular in many places. Besides France and the United States, modern district heating systems now warm homes in Iceland, Turkey, Poland and Hungary. The world's largest geothermal district heating system is in Reykjavik, Iceland, where almost all the buildings use geothermal heat. The air around Reykjavik was once very polluted by emissions from reliance on fossil fuels. Since it started using geothermal energy, Reykjavik has become one of the cleanest cities in the world.

Geothermal Heat Pumps

Another geothermal technology that helps keep indoor temperatures comfortable by using Earth's heat is the geo-exchange system, or geothermal heat pump. Geothermal heat pumps do not use geothermal reservoirs, so they can be used almost everywhere in the world—in areas with normal as well as high temperature gradients. By pumping fluid through loops of pipe buried underground next to a building, these systems take advantage of the relatively constant temperature 7–13°C (45–55°F) of the Earth right beneath our feet to transfer heat into buildings in winter and out of them in summer.

Geothermal heat pumps reduce electricity use 30–60% compared with traditional heating and cooling systems, because the electricity which powers them is used only to move heat, not to produce it. There are about 300,000 heat pump installations in the United States; Switzerland and several other countries are implementing heat pump programs. The U.S. Environmental Protection Agency rates geothermal heat pumps among the most efficient of heating and cooling technologies.

Direct Use Developments Worldwide

Geothermal direct use applications provide about 10,000 thermal megawatts (MW-th) of energy in about 35 countries. (In an additional 40 countries there are hot springs used for bathing, but facilities for commercial use have not been developed.) In the United States alone, there are some 18 district heating systems, 38 greenhouse complexes, 28 fish farms, 12 industrial plants, and 218 spas that use geothermal waters to provide heat.

Nearly every country has some areas underlain by low- and/or moderate-temperature geothermal waters. Expansion of direct uses of lower-temperature geothermal water can contribute to meeting the developing world's energy needs.

Renewability and Sustainability

Earth's heat is continuously radiated from within, and each year rainfall and snowmelt supply new water to geothermal reservoirs. Production from individual geothermal fields can be sustained for decades and perhaps centuries. The U.S. Department of Energy classifies geothermal energy as renewable.

Conservation of Resources

When we use renewable geothermal energy for direct use or for producing electricity, we conserve exhaustible and more polluting resources like fossil fuels and uranium (nuclear energy). Installed geothermal electricity generation capacity around the world is equivalent to the output of about 10 nuclear plants.

Worldwide, direct uses of geothermal water avoids the combustion of fossil fuels equivalent to burning of 830 million gallons of oil or 4.4 million tons of coal per year. Worldwide electrical production from geothermal reservoirs avoids the combustion of 5.4 billion gallons of oil or 28.3 million tons of coal.

Protection of the Environment

With all sources of energy, developers and consumers must work to protect the environment. The challenges differ with the type of energy resource, and the differences give geothermal energy certain advantages. Geothermal direct use facilities have minimal or no negative impacts on the environment. Geothermal power plants are relatively easy on the environment. They are successfully operated in the middle of crops, in sensitive desert environments and in forested recreation areas.

Protection of the Air and Atmosphere

Hydrogen sulphide gas (H_2S) sometimes occurs in geothermal reservoirs. H_2S has a distinctive rotten egg smell that can be detected by the most sensitive sensors (our noses) at very low concentrations (a few parts per billion). It is subject to regulatory controls for worker safety because it can be toxic at high concentrations. Equipment for scrubbing H_2S from geothermal steam removes 99% of this gas.

Carbon dioxide (a major climate change gas) occurs naturally in geothermal steam but geothermal plants release amounts less than 4% of that released by fossil fuel plants. And there are no emissions at all when closed-cycle (binary) technology is used.

Protection of Groundwater

Geothermal water contains higher concentrations of dissolved minerals than water from cold groundwater aquifers. In geothermal wells, pipe or casing (usually several layers) is cemented into the ground to prevent the mixing of geothermal water with other groundwater.

When highly-mineralized geothermal water needs to be stored at the surface, such as during well testing, it is kept in lined, impermeable sumps. After use, the geothermal water is either evaporated or injected back to its deep reservoir, again through sealed piping.

Visual Protection

No power plant or drill rig is as lovely as a natural landscape, so smaller is better. A geothermal plant sits right on top of its fuel source: no additional land is needed such as for mining coal or for transporting oil or gas. When geothermal power plants and drill rigs are located in scenic areas, mitigation measures are implemented to reduce intrusion on the visual landscape. Some geothermal power plants use special air cooling technology which eliminates even the plumes of water vapor from cooling towers and reduces a plant profile to as little as 24 feet in height.

By observing federal and state regulations, geothermal developers avoid interference with geysers and hot springs in areas set aside for their scenic beauty. Development in National Parks such as Yellowstone is specifically prohibited.

Improving Geothermal Technology

Since the 1970's the geothermal industry, with the assistance of government research funding, has overcome many technical drilling and power plant problems. Improvements in treatment of geothermal water have overcome early problems of corrosion and scaling of pipes. Methods have been developed to remove silica from high-silica reservoirs. In some plants silica is being put to use making concrete, and H_2S is converted to sulphur and sold. At power plants in the Imperial Valley of California, a facility is being constructed to extract zinc from the geothermal water for commercial sale.

As a result of government-assisted research and industry experience, the cost of generating geothermal power has decreased by 25% over the past two decades. Research is currently underway to further improve exploration, drilling, reservoir, power plant and environmental technologies. Enhancing the recoverability of Earth's heat is an important area of ongoing research.

Enhanced Geothermal Systems

Geothermal energy is accessible if there is sufficient heat, permeability, and water in a system, and if the system is not too deep. The available heat cannot be increased, but the permeability and water content can be enhanced. Private and government research projects in the United States, Japan and in Europe are improving the accessibility of geothermal energy by developing new technology to increase the permeability of the rocks and to supplement the water in hot, water-deficient rocks. Engineers estimate that by the year 2020, man-made geothermal reservoirs could be supplying 5 to 10% of the world's electricity.

Enhancing Reservoir Water

One unique example of enhancing reservoir water is at The Geysers steam field in California, where treated wastewater from nearby communities is being piped to the steamfield and injected into the reservoir to be heated. This increases the amount of steam available to produce electricity. With this enhancement, reservoir life is increased while providing nearby cities with an environmentally safe method of wastewater disposal.

Enhancing Reservoir Permeability

Permeability can be created in hot rocks by hydraulic fracturing—injecting large volumes of water into a well at a pressure high enough to break the rocks. The artificial fracture system is mapped by seismic methods as it forms, and a second well is drilled to intersect the fracture system. Cold water can then be pumped down one well and hot water taken from the second well for use in a geothermal plant. This "hot dry rock" technology is being tested in Japan, Germany, France, England and the United States.

The Future for Geothermal Energy

The outlook for geothermal energy use depends on at least three factors: the demand for energy in general; the inventory of available geothermal resources; and the competitive position of geothermal among other energy sources.

The Demand for energy will continue to grow. Economies are expanding, populations are increasing (over 2 billion people still do not have electricity), and energy-intensive technologies are spreading. All mean greater demand for energy. At the same time, there is growing global recognition of the environmental impacts of energy production and use from fossil fuel and nuclear resources. Public polls repeatedly show that most people prefer a policy of support for renewable energy.

The Inventory of accessible geothermal energy is sizable.

Using current technology geothermal energy from already-identified reservoirs can contribute as much as 10% of the United States' energy supply. And with more exploration, the inventory can become larger. The entire world resource base of geothermal energy has been calculated in government surveys to be larger than the resource bases of coal, oil, gas and uranium combined. The geothermal resource base becomes more available as methods and technologies for accessing it are improved through research and experience.

The Competitive Position depends primarily on cost:

Costs: Shorter and Longer Term

Production of fossil fuels (oil, natural gas and coal) are a relative bargain in the short term. Like many renewable resources, geothermal resources need relatively high initial investments to access the heat, hot water and steam. But the geothermal "fuel" cost is predictable and stable. Fossil fuel supplies will increase in cost as reserves are exhausted. Fossil fuel supplies can be interrupted political disputes abroad. Renewable geothermal energy is a better long term investment.

Costs: Direct and Indirect

The monetary price we pay to our natural gas and electricity suppliers, and at the gas pump, is our direct cost for the energy we use. But the use of energy also has indirect or *external* costs that are imposed on society. Examples are the huge costs of global climate change; the health effects from ground level pollution of the air; future effects of pollution of water and land; military expenditures to protect petroleum sources and supply routes; and costs of safely storing radioactive waste for generations. Geothermal energy can already com pete with the direct costs of conventional fuels in some locations and is a clean, indigenous, renewable resource without hidden external costs. Public polls reveal that customers are willing to pay a little more for energy from renewable resources such as geothermal energy

Costs: Domestic and Importing

Investment in the use of domestic, indigenous, renewable energy resources like geothermal energy provides jobs, expands the regional and national economies, and avoids the export of money to import fuels.

Energy demand is increasing rapidly worldwide. Some energy and environmental experts predict that the growth of electricity production and direct uses of geothermal energy will be revitalized by international commitments to reduce carbon dioxide emissions to avert global climate change and by the opening of markets to competition.

"Clean energy from the sea will be especially popular when the price of oil is no longer governed by economics but by environmental costs."

The Power of the Oceans Can Provide Clean Energy

France Bequette

France Bequette explains in the following viewpoint that new technology is helping scientists develop ways to utilize ocean energy. For example, tidal plants built at the mouths of estuaries use tidal fluctuations to drive turbines that generate electricity. Another way to transform the sea's energy into electricity, according to Bequette, is to build chambers that capture wave energy. Finally, ocean thermal energy—created by the difference between the surface temperature of tropical waters and water found at depths of four-thousand meters—can also be used to generate electricity. France Bequette writes for the *Unesco Courier*.

As you read, consider the following questions:
1. When did the first tidal power plant begin operating, according to Bequette?
2. According to the author, what was OSPREY?
3. What is a "tapering channel," according to Bequette?

The world's oceans could provide a limitless source of energy, according to Indian chemist Madanjeet Singh, an international authority on the subject. But exploiting ocean energy is no simple matter, if only because the possibilities of doing so differ from region to region. Experiments have been carried out with tidal and wave power and with ocean thermal energy. Few applications of these technologies are currently in use, but their number could increase in the coming centuries—if they attract investment on a massive scale.

A Clean and Inexhaustible Energy Source

Ocean tides constitute a clean and inexhaustible energy source, free from the climatic irregularities which are a constraint on wind and solar power. But places suitable for exploiting them are few and far between, for if a tidal power plant is to operate efficiently certain conditions must be met. A river estuary where the difference between high and low tide is at least five metres is necessary. It must be possible to construct a dam, and there must be a nearby source of electricity supply to make up for the intermittence of power production linked to the times of high and low tides.

One of the world's most suitable sites is the estuary of the river Rance, in western France, where the difference between high and low tides averages 8.17 metres, peaking at 13.5 metres during the equinoxes. The world's first tidal power plant began operating there in 1966. It is still the biggest, with a capacity of 240 megawatts (MW).

Tidal power plants consist of a high-capacity dam built across an estuary to hold back the water at high tide. At low tide sluice gates in the dam are opened to release a cascade of water that drives a turbine to generate electricity.

In China, says Singh, "eight plants with a total capacity of 6,210 kilowatts exploit tidal energy." There is a 20MW plant at Annapolis in Canada. But although many other suitable sites exist, construction costs are considered too high, especially as hydroelectric power is plentiful and cheaper. Tidal power plants are planned on Russia's White Sea and on the Severn and Mersey estuaries in the United Kingdom (UK).

In 1945, Japan became the first country to consider using sea waves as an energy source, followed by Norway and the

UK. The first power plant to use wave power, OSPREY (Ocean Swell Powered Renewable Energy), began operating in northern Scotland at the beginning of August 1995. A 2MW facility, OSPREY was designed along the following lines: waves entering a kind of submerged chamber open at the base pushed air into turbines to generate electricity sent via an underwater cable to the shore about 300 metres away. Unfortunately, the plant was damaged by the waves and then destroyed by a storm. The engineers who designed it did not give up, however, and a cheaper and more efficient version is being developed to supply small islands with much-needed electricity and to power a seawater desalinization plant.

Tidal Physics

The interaction of the Moon and the Earth results in the oceans bulging out towards the Moon, whilst on the opposite side the gravitational effect is partly shielded by the Earth resulting in a slightly smaller interaction and the oceans on that side bulge out away from the Moon, due to centrifugal forces. This is known as the Lunar Tide. This is complicated by the gravitational interaction of the Sun which results in the same effect of bulging towards and away from the Sun on facing and opposing sides of the Earth. This is known as the Solar Tide.

Australian Greenhouse Office, http://renewable.greenhouse.gov.au/technologies/ocean/tidal/html, June 1999.

Quite different technology is needed to turn the sea's thermal energy into electricity. The sea's surface temperature in the tropics ranges from 27 to 31 [degrees] Celsius (C) [81 to 88 degrees Fahrenheit] all the year round and that of deep water ranges from 6 [degrees] C at a depth of 1,000 metres to almost 0 [degrees] C [32 degrees Fahrenheit] at 4,000 metres. This temperature difference can be used to power a motor based on the same principle as that of the heat pump. A liquid is turned into gas in an evaporator, and then drives a turbine which generates electricity, before passing through a condenser, where it is turned back into liquid.

The snag is that the process requires huge turbines. The first practical application of this method was by the French engineer Georges Claude, who in 1930 loaded pipes and turbines onto a ship off the coast of Cuba. American engineers are working on the idea of siting floating power plants off the southern coast of the U.S. and one facility has been operating in Hawaii since 1981. Such plants could supply floating cities of the future with electricity, air-conditioning and fresh water, while the unpolluted, nutrient-rich cold water brought up from the depths could be used to breed fish and shellfish and grow edible seaweed.

But according to Scottish engineer S.H. Salter, the most promising device is a "tapering channel", invented by a Norwegian, Even Mehlum. As Singh describes it, "The waves are funnelled into a tapered natural or artificial channel. The water level rises and the water eventually spills into an elevated reservoir behind the narrow end of the channel. The water then flows back to the sea through a turbine, generating electricity in the process." This reliable, low-cost system is already operating in Norway and Java.

Limited Only by Imagination

Salter believes that if we use our imagination we will find an endless variety of ways of harnessing the sea's energy. One of them would be to use the difference between sea levels. In Egypt, water running through an underground canal linking the Mediterranean to the El-Qattara depression could be used to generate electricity. In Israel, the same principle could be used in a canal between the Mediterranean and the Dead Sea which would gradually descend 400 metres, although the estimated billion-dollar price tag is a deterrent.

Mr. Singh concludes that "clean energy from the sea will be especially popular when the price of oil is no longer governed by economics but by environmental costs. Cutting carbon dioxide emissions will help save the planet."

Periodical Bibliography

The following articles have been selected to supplement the diverse views presented in this chapter. Addresses are provided for periodicals not indexed in the *Readers' Guide to Periodical Literature*, the *Alternative Press Index*, the *Social Sciences Index*, or the *Index to Legal Periodicals and Books*.

Lester R. Brown	"Farmers Harvest the Wind," *Futurist*, November 2000.
David Case	"Cleaner Than Cows," *Progressive Populist*, March 15, 2001.
Seth Dunn	"The Hydrogen Experiment," *World Watch*, November/December 2000.
Christopher Flavin	"Bull Market in Wind Energy," *World Watch*, March/April 1999.
Christopher Flavin and Molly O'Meara	"Solar Power Markets Bloom," *World Watch*, September/October 1998.
Benjamin Fulford	"Here Comes the Sun: After Years of Cloudiness the Forecast for Solar Power Is Brightening," *Forbes*, February 5, 2001.
Scott Gold	"Green Energy Is Getting Its Second Wind," *Los Angeles Times*, September 12, 2000.
Greenpeace Magazine	"Here Comes the Sun," Summer 1998.
Bilal U. Haq	"Methane in the Deep Blue Sea," *Science*, July 23, 1999.
Douglas Jehl	"Curse of the Wind Turns to Farmers' Blessing," *New York Times*, November 26, 2000.
Roy McAlister	"Tapping Energy from Solar Hydrogen," *World & I*, February 1999.
Matt Scanlon	"Not Your Mother's Solar Power Anymore," *Mother Earth News*, January 2001.
Peter N. Spotts	"Will Ocean's Ice Crystals Yield Energy?" *Christian Science Monitor*, February 4, 1997.
U.S. Department of Energy	"Environmental and Economic Impacts of Geothermal Energy," www.eren.doe.gov/geothermal/geoimpacts.html, April 2001.
Hugh Westrup	"What a Gas!" *Current Science*, April 6, 2001.

CHAPTER 4

Should Alternatives to Gasoline-Powered Vehicles Be Pursued?

Chapter Preface

In Japan fifteen percent of all commuters ride bicycles to work. Fifty percent of routine travel in the Netherlands is done on bicycles. The city of Palo Alto, California has spent roughly one million dollars on building bicycle lockers and racks, bike bridges, and lighted cycle paths to promote commuting by bicycle. In fact, many governments around the world are encouraging people to ride bicycles as a way to lessen the use of gasoline-powered vehicles.

Critics of the automobile claim that driving motor vehicles creates more air pollution than any other human activity. As Marcia D. Lowe, a senior researcher at the Worldwatch Institute, points out, "Gasoline and diesel engines emit a large percentage of the carbon monoxide, hydrocarbons, and nitrogen oxides that result from fossil fuel combustion worldwide." Another problem with cars is that the gasoline they burn is made from oil, which many nations such as the United States must import. Many experts contend that this dependence on foreign oil suppliers makes a country like the United States vulnerable to sudden fluctuations in the supply and price of oil.

Not everyone believes that gasoline-powered vehicles are a serious problem, however. Many commentators point out that cars provide people with quick and comfortable transportation to work and school. Some, such as Marc Ross, a professor of physics at the University of Michigan, claim that as automotive technology improves, cars will have better fuel efficiency and emit fewer emissions in the future. Ross argues, "Over the past decade, manufacturers have made extraordinary advances in automotive technology." Other experts contend that technological advances will enable scientists to extract more oil from the earth, which will allow the automobile to remain a viable method of transportation for years to come. Moreover, many car advocates claim that alternatives to gasoline-powered vehicles such as bicycles do not provide the same travel-range or convenience as automobiles do.

Nevertheless, many governments worldwide are encouraging people to use bicycles instead of cars. Ellen Fletcher, who was a Palo Alto town council member in 1990 and is an advocate of the city's experiment in alternative transportation, claims that governments can convince people to pedal rather than drive. "All you have to do," she says, "is make it easier to ride a bike than drive a car. People will take it from there." The authors in the following chapter debate whether alternatives to gasoline-powered vehicles should be pursued. Taxing gasoline, providing subsidies for the development of alternative fuels, or constructing bicycle-only roadways are all ways that governments are encouraging the use of alternatives to gasoline-powered vehicles.

> *"The need for a viable alternative to the internal combustion engine is pressing. The world uses more petroleum now than at any other time . . . and the oil won't last forever."*

Alternative Vehicles Should Replace Gasoline-Powered Cars

Danylo Hawaleshka

In the following viewpoint Danylo Hawaleshka claims that it is important to develop alternative vehicles because the oil used to produce gasoline for traditional cars is running out. He explains that gas-electric hybrids and natural gas vehicles are currently being used to cut oil consumption, and in the future, fuel-cell vehicles may provide a permanent solution to the oil shortage. In addition, Hawaleshka reports that alternative vehicles can cut the emission of air pollutants. Danylo Hawaleshka writes for *Maclean's* magazine.

As you read, consider the following questions:

1. How much will the Durango SUV hybrid improve fuel economy over the gas-powered version, according to DaimlerChrysler?
2. According to Hawaleshka, how many automobiles will be in use worldwide in 2020?
3. What factors might limit the use of fuel-cell vehicles in the near future, according to the author?

The car is a showstopper, no doubt about that. Ever since Nigel Fitzpatrick got his red Honda Insight in May 2000, Vancouverites have been making little comments like "neat," or stopping and asking him how he likes driving a car with a gas-electric hybrid engine. His teenage daughter, in a testament to the Insight's cool factor, looks forward to being driven to school in it. Fitzpatrick, whose company, Azure Dynamics Inc., is developing a control system for a hybrid van, says he loves the sporty two-seater's tandem of a one-litre, three-cylinder engine seamlessly married to a quiet battery-powered motor. Visits to gas stations are infrequent. The only problem is the attention. "If you get on a ferry," says Fitzpatrick, "you've got to make sure you get out swiftly—otherwise, you're surrounded."

Solutions to Global Warming

The Honda Insight, as well as Toyota's four-seat Prius sedan, are just the first in a wave of alternative modes of transportation that industry analysts expect to see on North American roads over the next decade. Ford will offer its Escape sport-utility hybrid in 2003, and General Motors expects to release its full-size Chevrolet Silverado and GMC Sierra pickups in 2004, improving fuel economy by nearly 15 per cent. DaimlerChrysler is developing its Durango SUV hybrid for 2003, saying it will boast 20-per-cent better mileage than the gas-powered version.

The major manufacturers are also developing cars powered exclusively by batteries, like GM's speedy EV1, available through lease in California and Arizona. And they are eyeing natural gas, as well as the holy grail of automotive propulsion: non-polluting fuel cells. Meanwhile, a slew of futuristic bicycles propelled by small but powerful electric motors hold out the promise of alleviating, if only a little, the congestion in Canadian cities—and lungs.

The need for a viable alternative to the internal combustion engine is pressing. The world uses more petroleum now than at any other time in its history—about 75 million barrels daily—and the oil won't last forever. Gasoline prices, as every motorist knows, have been hitting record levels in 2000. Vehicular emissions play a major role in global warming, which most scientists believe is real and getting worse. Today, there are 700 million cars and trucks in use worldwide, and the number is expected to climb to one billion by 2020, according to Hiroyuki Watanabe, managing director of Toyota Motor Co. "The age of the internal combustion engine," says Watanabe, "is over."

That may be overstating things a bit, by current thinking anyway. The U.S. department of energy estimates hybrids will account for less than seven per cent of the world's vehicles by 2020, while electric- and natural-gas-powered vehicles will represent at most two per cent. Vehicles with fuel cells under the hood are expected to make up less than one per cent. Still, Watanabe's enthusiasm is understandable. Hybrids, unlike electric or fuel-cell cars, meet three key criteria for mass acceptance: reasonable price, range unhindered by spent batteries and convenient refuelling.

New Fuels

Honda's Insight gets up to 25 km per litre [68 miles per gallon] on the highway, about 10 km per litre [25 miles per gallon] more than its gasoline-fuelled equivalent, and retails for $28,820. Its batteries recharge under braking, harnessing the energy in its wheels. Honda has sold 3,000 Insights in North America, currently available only with a manual transmission, though an automatic is due in mid-2001. Toyota's Prius has the advantage of being a four-seater with an automatic transmission. Worldwide, Toyota has sold 40,000 Priuses, retailing in Canada for $29,990. Jean-Francois Banville, an Environment Canada engineer in Montreal, says hybrids will serve as a bridge between the comparative gas guzzlers of today and the fuel-sipping innovations of tomorrow, whatever those turn out to be. "Hybrids," says Banville, "will dominate the next 10 to 15 years."

Then what? Experts aren't quite sure. The industry is pushing hard to develop fuel-cell technology, which mixes compressed hydrogen with oxygen to produce electric power with no harmful emissions (or modest ones, if the hydrogen is first refined onboard from natural gas, methanol or petroleum). Burnaby, British Columbia (B.C.)-based Ballard Power Systems Inc. is at the forefront of fuel-cell development, and says DaimlerChrysler, Ford, General Motors, Toyota and Honda plan to have fuel-cell-powered cars available as early as 2004. But the prospect for their widespread use in the near term appears limited, due to their high cost and a lack of safe sources of hydrogen.

© Ed Stein. Reprinted by permission of Rocky Mountain News.

Natural gas offers an immediate alternative. John Lyon, president of FuelMaker Corp. of Toronto, estimates there are about 50,000 natural-gas-powered cars and light trucks operating in Canada. Compared with gasoline and propane, natural gas burns cleaner. Last month, American Honda Motor Co. Inc. took a 20-per-cent stake in Lyon's firm, which builds heavy-duty natural-gas compressors to refuel forklifts, fleet vehicles and Zambonis. With Honda's cash infusion, FuelMaker plans to shrink its patented technology and develop a $1,500 refuelling station for the home over the next three years.

The pump, says Lyon, will be designed for mounting indoors on a garage wall, rather like the outlet for a natural-gas barbecue. It will be the key, he says, to developing the consumer market for cars like the natural-gas-powered

Honda Civic GX, sold in the United States for up to $6,000 more than its gasoline equivalent. Despite the higher initial price, Lyon says natural gas is an economical option because the fuel costs the equivalent of about 40 cents a litre, compared with the mid-70s for gasoline. "That's basically half the price to travel the same distance," says Lyon. "That's quite an incentive."

Souped-up Golf Carts

Part of the global effort to clear the air revolves around getting drivers to use electric cars on short trips within cities. Ford's battery-powered Think City, a tiny but roomy two-seat hatchback capable of a top speed of 90 km/h [56 miles per hour] and a range of 85 km [53 miles], is available in Europe, and Ford plans to bring it to North America in 2002. In August 2000, Ottawa issued new regulations making it legal to sell low-speed electric vehicles for road use in Canada. These cars, sometimes disparagingly referred to as souped-up golf carts, will be limited to maximum speeds of 40 km/h [25 miles per hour]. The provinces, which control vehicle registration and driver licensing, are now mulling how best to deal with the vehicles.

Dynasty Motorcar Corp. in Kelowna, B.C., isn't waiting. The company is gearing up production for the U.S. market under the IT banner, for "intelligent transport." The plant, with a planned capacity of 10,000 units per year starting in April, will first produce a four-door electric hatchback and a van, with a convertible planned for the summer. Barry Good, Dynasty's chairman, sees his new markets in gated communities, airports, university campuses and resorts. Good expects a fully loaded, four-door IT with four seats to retail for $15,500.

If that's still too rich, an electric bicycle might be the ultimate low-speed alternative. The federal government has drafted regulations to permit power-assisted electric bikes with a top speed of 32 km/h [20 miles per hour] on the road. Marc LaFontaine, owner of three Sportable bike shops in the Ottawa area, is eager to see the regulations come into effect, likely in spring. LaFontaine is the Canadian distributor of eBikes, which retail for $2,000 to $3,500. He has 50 people on a waiting list to buy the electric bikes once they are legal. "If we could get five or 10 per cent of the commuters off the road," says LaFontaine, "think of the ramifications environmentally." Everyone, it's true, would breathe easier.

| "*To see environmentalists sign on to the new wave of electric or hybrid cars without considering the effects of making them . . . and using them . . . is dismaying.*"

Alternative Vehicles Do Not Solve Environmental Problems

Jane Holtz Kay

Jane Holtz Kay is an architecture/planning critic for the *Nation* and author of the book, *Asphalt Nation: How the Automobile Took Over America and How We Can Take It Back*. Kay argues in the following viewpoint that cars that run on alternative fuels damage the environment just as gasoline-powered automobiles do. According to Kay, one reason alternative vehicles harm the environment is that they still require highways, which lead to urban sprawl. Kay also claims that alternative vehicles cannot solve environmental problems because manufacturing them uses up valuable resources and produces dangerous greenhouse gases, which many scientists believe cause global warming.

As you read, consider the following questions:
1. According to Kay, what percentage of the emissions that cause global warming can be attributed to motor vehicles?
2. According to the author, on what did the Atmosphere Alliance blame an increase of 3.4 percent in U.S. emissions?
3. What percentage of a car's consumption of energy and resources can be attributed to its production, according to Kay?

You don't need a weatherman to tell you that the whole earth has become the scorched earth. And you don't need a climate course to tell you that the temperature has become hot news. In the hottest decade of the millennium, "severe weather alerts" have become as constant as the calendar, and shots of pieces of the melting polar caps as large as Rhode Island sinking into the sea have caused even the automakers to desert the Climate Control Coalition.

The Complicity of the Car

But if weather scares have chilled us out and heated our consciousnesses, there is one thing that the fluctuating thermometer and rising tides don't record. And that's the complicity of the car. Whatever the assessment of the damage of the capricious climate, the political and financial barometers have yet to register this largest single contributor to global warming.

"Is your current car too closely related to the fossil fuel it burns?" asks an advertisement for a luxury automobile. You bet it is. By spewing as much as half the US emissions that fuel climate disarray, our stock of motor vehicles is not only "related" to rising temperatures and erratic weather but a parent of the problem. In just one instance, the Atmosphere Alliance, has blamed a sharp jump of 3.4 percent in US emissions—more than the total of most nations—on one automated energy hog, the sports utility vehicle.

Alternative Cars Are No Solution

Earth Day was the ultimate. As emissions rose in automobiles and the selling campaign spread across the world, Denis Hayes waxed lyrical for the "clean car" with nary a word about alternate transportation or planning curbs. No surprise that Honda Insight advertises its [gas-electric hybrid] car as "just what you and the planet have been waiting for" or that Prius declares [its gas-electric hybrid] "The new car for a new world." But to see environmentalists sign on to the new wave of electric or hybrid cars without considering the effects of making them (30 percent of the resource consumption) and using them (sprawl-producing highways, pavement that adds to heat buildup, etc.) is dismaying.

"Hydrogen fuel cells are the mantra," says John Deciccio, a mechanical engineer with the American Council for an Energy Efficient Economy. "I feel this hybrid stuff has pulled the wool over their eyes."

And who out there worries about the 200 million vehicles that no one bothers to clean. Clock the minutes: Every second these motor vehicles travel 60,000 miles, use 3000 gallons of petroleum products and add 60,000 pounds of carbon dioxide to the atmosphere. That's two-thirds of US carbon dioxide emissions.

Cars Exact a High Personal Cost

The top urban areas with the longest annual [commuting] delays per person:

1. Los Angeles, 56 hours
2. Atlanta, 53 hours
   Seattle, 53 hours
4. Houston, 50 hours
5. Dallas, 46 hours
   Washington, 46 hours
7. Austin, Texas, 45 hours
   Denver, 45 hours
9. St. Louis, 44 hours
10. Boston, 42 hours
    Miami, 42 hours
    Nashville, Tenn., 42 hours
    Orlando, Fla., 42 hours
    San Jose, 42 hours
    San Francisco-Oakland, 42 hours

Kelly Thorton, *The San Diego Union-Tribune*, May 8, 2001.

The surprise is that despite the role of the motor vehicle in making the weather gyrate like a Dow Jones graph, the total contribution of America's auto-dependency remains a dirty but hidden secret. For even as the emissions used as we drive to work or errands contribute five tons of carbon dioxide a year for every car, we ignore the better part of the exactions: The roads we build to serve the car, the fuel we extract, the industrial energy consumed in producing 15 million motor vehicles a year are enormous—and largely unrecorded.

Finally, there are the longtime uncharitable environmental damages of dirty air and water or the human cost of 41,500 lives lost a year, before we even come to the environmental cost of this global warming in, say, re-doing the New York subway flooded last summer or restoring the damage from floods and fires.

Sleight of Hand

Part of the reason for the social innumeracy on the true cost of global warming is also bureaucratic. Instead of slotting the car's plural contributions of CO_2 into the "transportation sector" of the EPA listings, their compendium ignores most of them. By thus tucking them into the "industrial" or "manufacturing" category in its formal notations, the agency inadvertently helps the motor vehicle's harm elude detection.

Not only the motor vehicle and the highway's promethean influence avoid attention, however. Add production to this list. The car's production alone accounts for almost one-third of its consumption of energy and resources, according to the Environment and Forecasting Institute in Heidelberg, Germany. Consider the parts of the motor vehicle—push buttons for its windows, air-conditioning units, computers to control the engines; the plastic, asphalt and aluminum industries—and you see burn, baby burn.

So it is that even as the Kyoto Accords [which the United States signed in 1997 but never ratified] encourage scientists to push for reducing the energy from carbon sources . . . even as the housing industry is coaxed to be energy efficient . . . even as other executives from utilities to manufacturing respond, if slowly, to the climate issue, the car and the pricey sprawl and energy consumption its roads augment escape a reckoning.

The 12,000 miles per person a year, growing in both trips and heft, are not only stoking the climate's troubles at home but abroad. America's King of the Road lifestyle encourages non-industrialized nations to follow the US model at an alarming rate. Consider the consequences of a billion Chinese buying cars fueled by climate-heating coal. Combined with America's role as car exporter, the US rage for the road makes our sermons on energy abstinence, not to mention environmental good citizenship, hypocritical. Preaching what we don't practice deep sixes leadership. We "play God with climate" in Worldwatch Institute's phrase, and the car is the devil in many of the global warming details.

"*Air problems have given a boost to the electric car, once the stuff of futurists but increasingly seen on metro highways and streets.*"

Electric Cars Can Reduce Air Pollution

Tom Barry

Tom Barry contends in the following viewpoint that electric cars do not pollute the atmosphere the way gasoline-powered vehicles do. In addition, Barry claims that electric cars are fast and require little maintenance. He explains that although electric cars are expensive, new technology and increased demand will lower cost and improve their traveling range. Tom Barry writes for *Georgia Trend*, a monthly magazine.

As you read, consider the following questions:
1. How many miles could an electric car travel in 1999 before needing to be recharged, according to Barry?
2. According to the author, how much did electric cars cost in 1999?
3. How fast can a modified EV1 go, according to Barry?

W hat was that silent-running car that flashed past you in the High Occupancy Vehicle (HOV) lane at 75 mph, with just the driver on board?

A Solution to Air Pollution

Atlanta's air problems have given a boost to the electric car, once the stuff of futurists but increasingly seen on metro highways and streets.

Georgia Power Company, under a federal mandate to buy alternative-fuel vehicles, is adding another 100 electric vehicles to its fleet of 137 in Georgia. (The parent Southern Company plans to go up to 400 overall within several years.)

Company employees use them in daily operations, under a pilot commuter program that will soon include 100 employees. Employees can lease a Ford Ranger truck for $150 a month or a General Motors EV1 sports car for $200.

Georgia Power also loans out the nonpolluting vehicles to Metro Atlanta governments, universities, the Metropolitan Atlanta Rapid Transit Authority (MARTA) and top business customers, to boost demand for alternatives to the internal combustion engine. One advantage: Lone motorists can use interstate HOV lanes if piloting an alternative-fuel car.

Project manager Gary Floyd believes electric cars will be much more common in a few years as technology improves and costs go down. "We're seeing a lot of interest from the public."

Drawbacks and Solutions

The main drawback to electric vehicles has been their limited range (60–70 miles without recharging), few recharging stations and relatively high cost.

Ideally, electric cars are used where short, frequent trips are the norm, such as for security and maintenance trips around a university campus. Georgia Tech, Zoo Atlanta, Perimeter Mall, Lenox square and city of Atlanta government have used them.

Floyd says new batteries expected out in the next two years will double the electric car's range to 120 miles before it needs a recharge, which costs $1 and may take several hours. By the end of the year, he says, Georgia Power's recharging network in Metro Atlanta will include 200 stations, up from around 30 today. Motorists will be able to recharge at malls, the airport and other locales.

Floyd also expects vehicle costs—ranging from the low $30,000s to the mid-$40,000s—to come down as technology improves and volume increases. As it is, most major car manufacturers already produce electric cars in limited numbers, mostly for California and Arizona markets. Only the Ford Ranger is now sold in Georgia, but Floyd hopes the Georgia Power program will spur manufacturers to market more electric vehicles here.

Increasing Energy Security

Electricity-powered vehicles are two to three times more energy efficient than their internal combustion counterparts. And given that transportation accounts for more than 65% of U.S. oil consumption and that more than half the oil consumed in the country is imported, a transition to electric transportation will increase national energy security.

Eric Heim, *EPRI Journal*, Summer 1999.

"The internal combustion engine will be around for a long time, and until the prices on electric vehicles get a little lower, you won't see much demand," he says. "But with the air problems we have today, people are starting to look at alternatives. And if more limitations are put on gasoline vehicles or gas taxes go up, they'll look even more."

The EV Smile

The first-time driver of an electric car usually is surprised, says Floyd. Most electric cars, which are slightly smaller than their gasoline cousins, have a top speed of around 80 mph, although a modified EV1 has raced up to 184 mph. Manufactured by General Motors, the EV1 comprises about one-half of the Georgia Power fleet. Drivers expect sluggish and get quick, for the 137-horsepower, 2,970-lb. EV1 can go from 0 to 60 mph in 8.5 seconds.

"We call their response the EV Smile," says Floyd. "People learn the car is quiet and smooth and that it accelerates a lot better than they ever expected."

If the price tag is high, electric cars have far lower maintenance costs than gas-powered vehicles. According to Georgia Power, fuel costs for an electric vehicle are $18.75 pcr 1,000 miles, versus $36 for a comparable gasoline-powered car. The state of Georgia offers a $1,500 tax credit for alternative-fuel vehicles, and there's a federal tax credit on the purchase of an electric car, up to $4,000.

Other than battery replacement after a few years (batteries today cost several thousand dollars) and checking brake and power steering levels, not much service is required.

Run silent, run fast, run clean.

> "Cars running on electricity may create no
> air pollution themselves, but the electricity
> has to come from somewhere to charge and
> re-charge the batteries that run these cars."

Electric Cars Do Not Reduce Air Pollution

Thomas Sowell

Thomas Sowell argues in the following viewpoint that electric cars are unsafe, impractical, and polluting. He contends that the design of electric cars is flawed: electric cars are so small that they are unsafe in highway collisions, and they run on batteries that need to be recharged too often to be practical for most transportation needs. Most important, Sowell contends that although electric cars themselves do not pollute, the electricity used to run them is generated at power plants that emit pollution. Thomas Sowell is a syndicated columnist.

As you read, consider the following questions:

1. What condition has California's government imposed on car manufacturers, according to Sowell?
2. In Sowell's opinion, who will inevitably pay for the electric cars that nobody wants?
3. Why doesn't the public want to buy electric cars, according to the author?

From "Californians Have No Sense of Proportion," by Thomas Sowell, *Conservative Chronicle*, February 7, 2001. Copyright © 2001 by the Creators Syndicate. Reprinted with permission.

Mathematicians use the term "rational numbers" for numbers that can form a ratio. By this definition, there is a lot of irrationality in California, where many people seem incapable of forming a ratio or proportion between different things.

California's 2001 electricity crisis is a result of years of refusing to have any sense of proportion between the desirability of environmental goals and the desirability of having electricity. Yet apparently the state's politicians have learned nothing from any of this.

Impossible Quotas

Having provoked an electricity crisis and a financial crisis by imposing impossible conditions on public utilities, the 2001 California government is imposing similarly irrational conditions on the automobile industry by requiring them to produce a certain quota of electric cars for sale in the state, as a precondition to their selling any other cars in California.

The purpose of the electric cars is to reduce the air pollution created by cars that burn gasoline. Obviously, no one is in favor of polluted air, but the question is whether the desirable goal of reducing pollution is to be pursued in utter disregard of other desirable things.

Electric cars may be fun at amusement parks, where they don't have to go very far or very fast. But if the consuming public wanted electric cars for regular use, Detroit would be manufacturing them by the millions. Only people infatuated with their own wonderful specialness would think that their job is to coerce both the manufacturers and the consuming public into something that neither of them wants.

California seems to have more than its fair share of self-infatuated people proclaiming utopian notions. Worse yet, such people are indulged by the media, the political system and the courts, while the enormous costs they create are quietly loaded onto unsuspecting consumers and taxpayers.

Somebody is going to have to pay for these electric cars that the public does not want. State agencies can buy some of them with the taxpayers' money. Some private individuals and organizations may be subjected to pressure from the state government to buy them. And some electric cars may just sit on dealers' lots or in storage, gathering dust. But they are still all going to have to be paid for by somebody because there is no free lunch.

Paying the Costs

Maybe those who imposed these new requirements think that the automobile companies can be forced to absorb the losses. Imposing costs on people out of state is a ploy that has been tried before with electricity. But apparently some people never learn.

Nothing is easier than glib enthusiasm for the benefits of electric cars—and some of those benefits may even be real. But there is still the need to have a sense of proportion, because there are other benefits that will have to be sacrificed and other costs that will have to be paid.

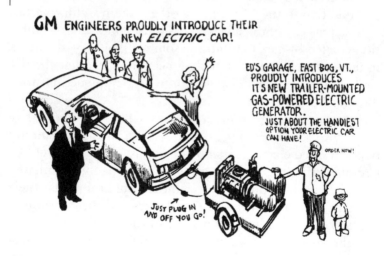

© Danziger. Reprinted by permission of the Christian Science Monitor.

Electric automobile engines are not powerful enough to move full-size cars at any reasonable speed, so that means people have to drive around in flimsy vehicles that can easily become death traps in an accident. Make no mistake about it, air pollution increases the incidence of fatal diseases. But will more people die from that than from traffic deaths in flimsy cars? People who are crusading for electric cars are not interested in that ratio.

Cars running on electricity may create no air pollution themselves, but the electricity has to come from somewhere to charge and re-charge the batteries that run these cars. What difference does it make if the car itself creates no pollution but the pollution occurs at an electric power plant, miles away, that is the ultimate source of the energy that moves the car?

Why doesn't the public want to buy electric cars? Because in real life you have to be able to get where you want to go, in some reasonable time, whether or not your destination is within the narrow range of an electric car's batteries. And you want to be able to turn around and come back when you are ready, not have to wait for hours to recharge your batteries for the return trip.

You may not get there at all if you are oozing down a highway in a fragile little vehicle that is out of sync with the fast-moving heavy traffic around you. But none of this matters to people who are not in the habit of weighing one thing against another. Nor do such people want to allow other people to weigh one thing against another for themselves, rather than have their choices dictated from on high. No sense of proportion.

"*The hydrogen fuel cell . . . is getting a fresh
look as possibly the most promising
replacement for the internal combustion
engine in the early part of the 21st
century.*"

Fuel Cell Cars Are Clean and Efficient

Donald W. Nauss

In the following viewpoint, Donald W. Nauss reports that automakers are developing an affordable car that will be powered by hydrogen fuel cells rather than gasoline. Because hydrogen is not yet conveniently sold at retail pumps the way gasoline is, DaimlerChrysler is designing a fuel-cell car that converts gas to hydrogen, although such a vehicle would still produce some harmful emissions. In the future, automakers will build hydrogen-powered autos that emit no harmful greenhouse gases. Donald W. Nauss is a staff writer for the *Los Angeles Times*.

As you read, consider the following questions:

1. According to Nauss, how much would a hydrogen infrastructure cost?
2. How much lower will be the emissions of Chrysler's hybrid vehicle compared to conventional cars, according to the author?
3. According to Nauss, what is the biggest obstacle to the success of fuel-cell cars?

As the search accelerates for better power sources for cars and trucks, a proven but problematic technology—the hydrogen fuel cell—is getting a fresh look as possibly the most promising replacement for the internal combustion engine in the early part of the 21st century.

Problems and Solutions

But fuel cell vehicles still face significant hurdles, particularly in finding room to store the hydrogen in the car and in the massive job of developing a network for getting hydrogen to motorists—an infrastructure that could cost up to $150 billion to put in place.

Now Chrysler claims it has come up with a practical solution to these problems, one it says could move the timetable forward 10 years—as early as 2007—for the introduction of significant numbers of affordable fuel cell vehicles.

Chrysler will announce in 1997 at the North American International Auto Show . . . that it has developed a way to convert conventional gasoline into hydrogen on board the vehicle. The hydrogen is fed into a fuel cell, then combined with oxygen to produce electricity to run the vehicle.

"With this technology, we think fuel cell vehicles can be here within 10 to 15 years," said Christopher Borroni-Bird, advanced technologies specialist for Chrysler.

Chrysler is not claiming any breakthrough in fuel cell technology, and other auto makers are also working on fuel cell vehicles—both Mercedes-Benz and Toyota introduced prototype vehicles in 1997. But Chrysler is the first to pursue one that produces hydrogen on board from existing fuel sources.

"This is a big step forward in transforming the kinds of cars we drive in the next century," said Frank O'Donnell, deputy director of Fuel Cells 2000, a nonprofit organization that promotes the technology.

The race to develop alternative-fuel vehicles has vast international implications. Auto makers and governments in Germany, Japan and the United States are pouring hundreds of millions of dollars into research of electric and hybrid vehicles powered by fuel cells, flywheels, ultracapacitors and exotic fuels.

In the United States, the Clinton administration has tried to kick-start Big Three [General Motors, Ford and Chrysler] research and development efforts through the Partnership for a New Generation of Vehicles (PNGV). The industry-government initiative is an effort to develop an affordable 80-mile-per-gallon family car by 2004.

Chrysler's proposed fuel cell vehicle is a byproduct of the PNGV effort. The company says it will produce a working model in two years, a hybrid vehicle that also uses batteries for cold starts and quick acceleration.

Safer than Conventional Cars

Actually an ultralight advanced-composite fuel cell car can be safer than a conventional car. The advanced-composite materials in a correctly designed crush structure can absorb five times as much crash energy per pound as steel, and do so more evenly so that the crush link or stroke can be used twice as effectively. It is possible, for example, to design—and Hypercar Inc., has designed—an SUV replacement Hypercar vehicle comparable in interior volume and superior in load hauling and other attributes to the most popular mid-sized sports utility vehicles (SUVs).

Carl J. Levesque, *Public Utilities Fortnightly*, February 15, 2001.

The auto maker claims its six-passenger sedan will be quieter and more durable than a conventional gasoline-powered car while matching its performance. It will be 50% more fuel-efficient than today's vehicles and have 90% lower emissions.

Environmentalists point out that other auto makers are developing fuel cell vehicles with no noxious emissions. They say Chrysler's approach might get fuel cell vehicles to market faster, but it will continue our reliance on imported oil by pushing back development of an integrated hydrogen production, distribution and refueling system.

"It could be the right technology but the wrong fuel," said Jason Mark, an energy analyst with the Union of Concerned Scientists. "It could be a stepping stone or a roadblock."

However, Chrysler, whose research and development budget is much smaller than those of domestic rivals General Motors and Ford, says it is taking a marketing approach toward the development of alternative-fuel vehicles. It says it wants to engineer technology to meet the needs of customers, not manipulate consumer taste to meet a new technology.

The auto maker maintains it is more practical and economical to use the existing gasoline-based system of production and retailing rather than rely on development of a hydrogen-based one that would take decades and tens of billions of dollars to build.

"Basic economics tell us this infrastructure is not going to change overnight just because car companies have fuel cell prototypes that run on hydrogen," said Francois Castaing, Chrysler vice president of vehicle engineering.

Fuel Cell Technology

Fuel cells were invented in 1839 by Sir William Grove but remained largely a curiosity until National Aeronautics and Space Administration (NASA) made extensive use of them in space missions. Both the Gemini and Apollo spacecraft used fuel cells.

They operate on a basic electrochemical reaction that scientists have long understood—essentially the reverse of electrolysis. In electrolysis, an electric current is run through water to break the liquid down into its basic components, hydrogen and oxygen. A fuel cell, using a platinum catalyst, combines hydrogen with oxygen to create electricity and water. In vehicles, the electricity is routed to small motors in the wheels.

For autos, the attraction of fuel cells is that they combine some of the advantages of conventional gas cars—long range and quick acceleration—with the quiet, low maintenance and clean emissions of electric vehicles.

The use and development of fuel cells for cars and trucks has been held back by their cost, size, performance and storage and infrastructure issues. But great strides have been made in some of these areas in recent years.

For instance, a decade ago a fuel cell cost 1,000 times more than a gasoline-powered engine. Today, it costs about $30,000, only 10 times that of a typical internal combustion engine.

"The cost is coming down dramatically and we suspect it will continue to come down," said Peter Rosenfeld, Chrysler's director of advanced technical planning.

The size of fuel cells has also been reduced greatly. In

1994, when Mercedes-Benz unveiled a fuel-cell-powered prototype minivan, the fuel cell occupied the entire cargo bay, leaving room for only a driver and a passenger. In April 1996, it showed an updated version of the same vehicle with the fuel cell packed under the van's rear seats—and room for a driver and six passengers.

Already, fuel cells are beginning to show up in some large vehicles. Ballard Power Systems of Vancouver, Canada, is building three fuel cell buses that will be on the streets of Chicago in 1998. It is also supplying Mercedes-Benz with fuel cells.

Storage and Safety

Hydrogen, a gas when at normal temperatures, can be compressed and stored in heavy metal cylinders. They tend to be bulky, capable of carrying only enough of the gas to provide about half the range of traditional cars. Hydrogen can also be liquefied at low temperatures or stored in a solid state using expensive metal hydrides or alloys. Toyota is using a form of solid hydrogen in its prototype shown in October 1997.

Safety is also a primary concern with hydrogen, which is volatile and highly flammable as a gas. But most experts consider hydrogen no more dangerous than gasoline or natural gas, and they say it can be safely stored on board vehicles.

An even bigger obstacle than storage is the almost total absence of a hydrogen production, distribution and refueling infrastructure that will be essential if such vehicles are to become practical for motorists.

"We've seen very little success in developing any alternative-fuel infrastructure," Mark said.

To address this issue, a number of companies have been working to develop small processors, or so-called reformers, that turn more traditional fuels such as gasoline into hydrogen and other harmless gases on board the vehicle. Chrysler is using technology developed by the laboratory division of the Arthur D. Little Co. consulting company in Cambridge, Mass.

"Getting alternative fuels started is always the hardest part because of the lack of a fuel infrastructure," said Jeffrey Bentley, director of technology and product development for Little. "We have tried to leapfrog the infrastructure issue by using gas right out of the pump."

Fuel-on-Board

The processor consists of a number of units that break down the gasoline into hydrogen for the fuel cell. First, a vaporizer—6 inches in diameter and 20 inches long—converts the gasoline to a gaseous form.

Next the gas moves to a partial oxidizer, a 14-by-22-inch canister equipped with a sparkplug to initiate partial burning. The vaporized fuel is then combined with air to produce hydrogen and carbon monoxide.

The gases move into another unit where steam acts with a catalyst to turn the harmful carbon monoxide into carbon dioxide. Additional hydrogen is also produced from this process.

Finally, the fuel undergoes oxidation in which injected air reacts with a catalyst to remove any remaining carbon monoxide, leaving hydrogen-rich gases that are fed into the fuel cell.

Chrysler's six-passenger sedan is likely to have a tailpipe for emission of carbon dioxide, nitrogen and water, most of which will be reused by the on-board processor.

Though the vehicle is not zero-emission, Chrysler hopes it might eventually qualify as an equivalent zero-emission vehicle, a new class of low-emission autos under consideration by the California Air Resources Board.

By bringing a practical fuel cell vehicle to the market quicker, Chrysler argues, the clean air benefits will be greater than if it were required to wait to make zero-emission vehicles that rely on a hydrogen infrastructure.

"This is not pie in the sky," Rosenfeld said. "This is a potential alternative to the engine of today."

"*[Natural Gas Vehicles] can reduce carbon dioxide, the principle 'greenhouse' gas, emissions by almost 20%.*"

Natural Gas Vehicles Can Reduce Greenhouse Gases

Steve Spaulding

Steve Spaulding contends in the following viewpoint that cars that run on natural gas run cleaner than conventional cars and therefore reduce exhaust emissions, including greenhouse gases. In addition, he claims that since the United States has abundant natural gas resources, driving natural gas vehicles rather than conventional cars could decrease U.S. dependence on foreign oil suppliers. Spaulding argues that natural gas vehicles are safe, economical, and require little maintenance. Steve Spaulding is the production editor for *Contractor* magazine.

As you read, consider the following questions:
1. According to Spaulding, what percentage of the U.S. supply of natural gas is produced domestically?
2. How many natural gas vehicles were being driven in 1997, according to the author?
3. According to Spaulding, how much lower was the injury rate for natural gas fleet vehicles compared to gasoline fleet vehicles in 1992?

E ight years ago, Jack Fiora, the owner of Fiora Electrical Construction in Meriden, Conn., could see the writing on the wall. The '80s construction boom in New England had been good for local contractors, but the economy was leveling out and work was beginning to dry up. Fiora began looking for ways to reinvest his profits back into the 50-year-old business that would make for a more efficient operation and long-term savings. It was one of his customers that led him to use natural gas to power his fleet of trucks, and he's been thankful ever since.

A Cheap Domestic Resource

"Back in 1989 we were busy. We were spending between $5,000 and $6,000 a month in gasoline," Fiora says. "Yankee Gas (the local utility) was one of our customers, and one day they sent me a flyer out of the blue. I called them up on a whim."

The immediate attraction was cost. [In 1997] natural gas cost about a third less than gasoline for the same amount of energy. In 1996, the average price for a gallon of gasoline was $1.29 compared with a natural gas average price of 78 cents for the energy equivalent.

Natural gas is an abundant, domestic natural resource, with about 88% of the U.S. supply produced in the United States (more than 50% of the oil used here is imported) meaning few tariff costs are passed on to consumers. Neither is the cost of refining; natural gas can be used almost straight out of the ground after a little mixing and balancing. Thanks to the home heating market, an extensive pipeline delivery system of 1.3 million miles is already in place throughout the country, keeping transportation costs low as well.

Clean-Burning

Another attraction for Fiora was that natural gas is the cleanest-burning of all the fossil fuels. The major component of natural gas is methane, which makes up more than 90% of "pipeline quality" natural gas. The main byproducts of combustion are carbon dioxide and water.

In the big picture, this makes for a cleaner environment. While Fiora admits to being "a little bit of an environmentalist," he was more interested in the longer engine life and lower maintenance costs that result from cleaner emissions.

Fiora decided to start with a test case to see how well the technology performed.

"We put it into my Jeep," he recalls. "I said: 'Look, we'll try it here. If it works, we'll start converting the trucks.'" He ended up converting all four of them.

"For the Jeep it was, at the time, two grand," Fiora says. "For our bigger trucks it was maybe $3,500. It matters on the tankage, because that's where most of the money goes. The electronics that you put under the hood aren't that bad, but they've got these new carbon-fiber tanks. Lightweight, but they're big bucks."

For all the money laid out, the system paid for itself in the first year of operation.

"At the time I was putting on 50,000 miles a year," he says. "Just add that up, take a third off your gas bill and it's paid for in no time."

And Fiora is still driving his Jeep, which is a particular point of pride. "This Jeep now has 300,000 miles on it, and I never touch that engine," he says. "The oil goes for 10,000 miles, and you'd never know it was dirty. I change it between eight and 10, but every time I go to Jiffy Lube or someplace they ask me why I'm bothering to change what looks like brand-new oil. Really, it's amazing what it does."

The Basics of NGVs

Until 1992, the only way to obtain a natural gas vehicle (NGV) was to convert an existing gasoline or diesel vehicle. Conversion is still a cost-effective and practical option for many NGV users, but now several engine and vehicle manufacturers such as GM, American Honda, John Deere and Ford produce and sell factory-built NGVs.

NGVs may be either bi-fuel or dedicated. Bi-fuel vehicles run on either natural gas or gasoline, with fuel selection being made either by the driver or through an automatic switching system. Dedicated NGVs run only on natural gas. Typically, dedicated NGVs display better performance and fuel efficiency because they are able to take full advantage of the characteristics of only one type of fuel.

There is little difference in price between bi-fuel and dedicated vehicles from original equipment manufacturers—a few hundred dollars either way, depending on the needs of the user. The price of converting varies too greatly from case to case for any practical price comparison of bi-fuel versus dedicated.

In bi-fuel conversions, special onboard fuel storage cylinders for the natural gas are added, along with electronics and a fuel management package. Dedicated NGV conversions also require the cylinders, electronics and fuel management package, but the gasoline tanks are either sealed or removed, and fuel management is simpler.

Open- and Closed-Loop Systems

Prior to 1985, most gasoline vehicles had carbureted engines, and conversion equipment for those vehicles was called open-loop. Open-loop systems were set to allow the engine to perform at idle and acceleration, but there was no mechanism to allow the conversion system to interact with the engine to enable it to self-adjust.

For today's fuel-injected systems, a conversion system called closed-loop is used. Closed-loop systems have mechanisms that work with the vehicle's computerized engine controls. Performance and emissions are continuously monitored and adjusted. While both open-loop and closed-loop systems are still available, closed-loop systems tend to give better performance. All factory-built NGVs use closed-loop systems.

Fiora has felt the difference between open-loop (with his Jeep) and closed-loop systems.

"The system I have, which is an older system, there is difference," Fiora says. "Say you're going up a hill. In a gasoline-powered vehicle you would have the pedal half down. In my Jeep, I have to put it three-quarters down. That's pretty much it. With the newer [closed-loop] systems I've driven there's no difference at all."

The cost premium of an NGV over a traditionally fueled vehicle can vary from $3,000 to $5,500 for a light-duty pickup or van, with roughly half that amount going toward on-board storage cylinders. Since initial cost is one of the biggest factors affecting the NGV market, cylinders have become one of the most intense areas of Research and Development (R & D) in the industry. The Gas Research Institute has a goal to reduce the cost of cylinders by 40% to 60% by the year 2000.

Practical Matters

About 60,000 natural gas vehicles are being driven in the United States in 1997. Most are used in commercial or government fleets and buses. According to the American Gas Association, the number of NGVs in use grew 78% from 23,192 to 41,227, between 1992 and 1994. AGA projections estimate that about 1 million NGVs could be on U.S. roads by 2015. Currently, one in five new bus orders are for NGVs.

There are 1,200 natural gas fueling stations around the country. Fueling with natural gas operates on the principle of equalizing pressures between the fueling station's storage tanks and the vehicle's on-board storage cylinders (which store at pressures from 2,400 psi to 3,600 psi).

Outselling Conventional Cars

Advocates say natural gas vehicles (NGVs) should be outselling conventional gasoline-fueled vehicles. NGVs produce lower emissions, various government initiatives can help defray their higher cost ($3,000-$5,000 higher than conventional vehicles), and they cost less to fuel—89 cents/gallon-equivalent of natural gas compared with an average of $1.23/gallon in 1999 for regular unleaded gasoline.

Christian Schmollinger, *Oil Daily*, November 4, 1999.

NGVs can be refueled either by quick-fill or timed-fill. Quick-fill is used when vehicles need to be refueled in a time period similar to that of gasoline vehicles, about three to seven minutes. At a quick-fill station, a compressor stores natural gas in a high-pressure tank. Refueling is done at the tank while the compressor replenishes the tank's supply. The system shuts off when the cylinder capacity of the vehicle is reached, or at whatever point the user desires.

A timed-fill system eliminates the need for a high-pressure storage tank at the station. Natural gas is pumped by the compressor directly into the on-board cylinders over the course of six to eight hours. Timed-fill is usually for fleet vehicles that return to a specific location at the end of a working day.

While natural gas stations are still few and far between compared to gasoline stations, natural gas suppliers are proving themselves more accommodating in meeting the needs of their users. If a home or business already has a natural gas hook-up (for heating, cooking, etc.) the local gas utility can install an on-site refueling system, making the user independent of outside filling stations. There are also a variety of portable fuel delivery systems that use over-the-road transportation of compressed natural gas in tube trailers. For interstate travel, the Natural Gas Vehicle Coalition offers a free booklet, "The Pocket Guide to NGV Fueling Stations in the United States," for planning trips.

Safety

NGVs have a remarkable safety record. A 1992 AGA survey of more than 8,000 fleet-based vehicles found that in 278.3 million miles, NGV injury rates per vehicle mile traveled were 37% lower than the rate for gasoline fleet vehicles, 34% lower than the rate for the entire population of registered gasoline vehicles and were without a single fatality (compared to 2.2 fatalities per 100 million miles for gasoline vehicles).

This safety record is due at least in part to the chemical properties of natural gas. Leaked natural gas dissipates into the atmosphere, instead of pooling on the ground. Its ignition temperature is 1,200[degrees] F, as opposed to 600[degrees] F for gasoline. It will burn only when the proper air-to-fuel ratio is reached—natural gas will not ignite when air concentration is below 5% or above 15%. Natural gas contains only trace amounts of toxic substances and is neither carcinogenic nor caustic. Another reason may be the stringent standards for on-board storage cylinders.

Government Initiatives

Susan Jacobs, a representative of the Natural Gas Vehicle Coalition (an industry group committed to developing long-term markets for NGVs) was an attendee at the 15th Annual Natural Gas Vehicle Conference held Sept. 7–9, 1997 in Salt Lake City.

"We had an attendance of over 750 and over 100 display booths," she says. "We had a new format this year with break-out sessions for different topics, like airport carriers, refuse haulers, taxi cabs and so forth. We also had sessions on federal legislation and state and local initiatives."

While every industry needs to keep track of what legislators are doing, the natural gas industry is in a particularly good position with respect to the government. Ever since the early 1990s, the feds have been throwing money and incentives at all levels of the industry. The two main reasons for this are that natural gas is friendly to the environment and can play a major role in reducing U.S. dependence on foreign oil.

Factory-built dedicated NGVs have the potential to reduce exhaust emissions of carbon monoxide by 70%, non-methane organic gas by 89%, and nitrogen oxides by 87%. In addition, they also can reduce carbon dioxide, the principle "greenhouse" gas, emissions by almost 20%. Because of this, natural gas qualifies as a "clean fuel" under the Clean Air Act Amendment of 1990.

Provisions of the amendment require fleet operators of 10 or more automobiles or light-duty trucks and vans that are capable of central fueling and are located in areas of poor air quality to begin purchasing clean-fuel vehicles by model year 1998.

Reducing Dependence on Foreign Oil

Another important piece of legislation is the Energy Policy Act of 1992, designed to establish a firm U.S. energy policy and to reduce dependence on foreign oil.

U.S. Department of Energy projections show oil consumption in the United States growing from about 17.4 million barrels per day to 20.7 million barrels by 2005. For each reduction in oil demand of 100,000 barrels per day in the year 2000, DOE estimates the annual oil import bill would fall by nearly $1 billion and, incidentally, create between 15,000 and 25,000 American jobs.

Provisions of the Energy Policy Act deal with alternative transportation fuels, requiring their use in certain fleets located in metropolitan areas with a 1980 population of 250,000 or more. In 1997, 33% of all vehicles in the federal fleet that fall under the act are running on alternative fuel, with that percentage rising to 75% by the year 2000.

Incentives vary from state to state, Jacobs says, with California leading the push. She suggested that contractors and other fleet truck owners get more information from their state legislators.

"There are also some federal dollars that are available in the form of tax incentives," Jacobs says. "There are incentives for putting in a fueling station, for the cost of conversion and for the outright purchase of fleet vehicles."

Jack Fiora got such a good deal from the state of Connecticut that he wonders, at times, why more people don't make the switch.

"A lot of people have taken notice since the switchover," he says. "Most are very interested, but I think it's still a mystery to a lot of people. Some people don't like to change."

For his own part, he's glad he made the move when he did. "What we did was an investment in our company. It really helped to keep us going through the bad years."

> "While . . . cheaper ethanol would only begin to reduce our perilous reliance on foreign oil, it would offer fringe benefits, such as lowering U.S. emissions that contribute to global warming."

Ethanol Is a Clean and Renewable Fuel

David Stipp

In the following viewpoint David Stipp argues that ethanol—a renewable energy source made from corn—burns cleaner than gasoline does. He claims that although ethanol may soon be cheaper to purchase than gasoline, it is difficult to obtain because few service stations sell it. Stipp contends that as scientists develop cheaper ways to produce ethanol, and oil companies begin selling more of the fuel, this renewable energy source could reduce U.S. reliance on foreign oil supplies. David Stipp writes for *Fortune* magazine.

As you read, consider the following questions:
1. What is E85, according to Stipp?
2. According to Stipp, what are flex-fuel vehicles?
3. What percentage fewer miles per gallon do vehicles running on ethanol get compared with cars running on gasoline, according to the author?

E thanol is emerging as the first "biofuel" that doesn't invite guffaws from Saudi sheikhs. Surging oil prices have brightened its outlook, as have federal clean-air rules favoring its use. But what really gives the Cornbelt brew potential is the progress made in lowering its cost. Within a few years ethanol could be cheaper at the pump than gasoline. Since the mid-1980s its cost has fallen about 25%. At scattered Midwest service stations, E85—an ethanol-based fuel that some newer pickups and other vehicles are equipped to handle—has recently sold for 15 to 25 cents less per gallon than regular unleaded gasoline.

Supply and Demand

There's a hitch, though: The oil industry owns the pumps, and for rational business reasons has long resisted corn-made ethanol and other "renewable" alternatives. Although more than a million U.S. cars and trucks produced since 1995 can run on E85, a blend of 15% gasoline and 85% ethanol, only about 100 gas stations in the U.S. offer it.

The gap between supply and potential demand is growing fast, since federal rules give carmakers extra mileage credits for "flex-fuel" vehicles that can handle ethanol fuels. The credits help the companies meet fuel-economy requirements set by Congress. By 2004, General Motors (GM) alone says that it will have produced more than a million flex-fuel trucks. These are equipped with sensors and computers that automatically adjust the engines' fuel-air mixture to accommodate varying ethanol content. Ford says it's putting about 250,000 flex-fuel vehicles, including its Taurus LX sedans, on the road each year.

Legislators are also rallying behind ethanol. California and other states are phasing out MTBE—a chemical now widely added to gasoline to cut carbon-monoxide emissions—because of leaks that have poisoned groundwater. A recently introduced U.S. Senate bill would ban MTBE additives in four years and boost the use of alternatives like ethanol. If it becomes law, demand for ethanol should triple over the next decade, says Trevor Guthmiller, executive director of the American Coalition for Ethanol in Sioux Falls, S.D.

Even if demand triples, ethanol would still represent only

about 3% to 4% of the U.S. fuel market. There are such large institutional and structural barriers to ethanol that its price may need to drop even further before it becomes a major bulk fuel. One reason is that it has less "energy content" than gasoline; cars running on E85 get 10% fewer miles per gallon than they would on gasoline. Another is that ethanol's current price is artificially low—federal tax credits typically knock about 46 cents a gallon off E85's manufacturing cost. (Ethanol producers note that gasoline prices are heavily subsidized by oil-industry tax breaks and other factors.) Perhaps most important, ethanol producers and gasoline suppliers would have to make major capital investments to put E85 in pumps nationwide—expenditures most of them are likely to deem too risky unless demand is ensured.

Soaking Up Greenhouse Gases

Fossil fuels release carbon into the air but ethanol is produced from agricultural crops that soak it up. This carbon cycle maintains the balance of carbon dioxide in the atmosphere. As a recent report by Argonne National Laboratory concluded, ethanol from corn reduces fossil energy use by 50-60% and greenhouse gases by 35–46% compared with conventional gasoline.

Eric Vaughn, *Chemistry and Industry*, May 3, 1999.

Bioengineers are working to slash production costs as much as possible. Most ethanol is now made by fermenting the starch in corn kernels. Cheaper, more plentiful feedstocks, such as corn stalks and leaves, could be used, however, with the help of "cellulase" enzymes to break down cellulose, the tough stuff that holds plants together. Cellulases aren't exotic—they're used to make stone-washed jeans—but they're too pricey for making ethanol. The U.S. National Renewable Energy Lab in Golden, Colo., recently launched a three-year, $17 million project with Genencor International, a Palo Alto biotech company, to develop cheap bioengineered cellulases. The project's goal is to cut the enzymes' costs by a factor of ten, which in turn could trim ethanol's cost by a fourth or more.

If that happens, irresistibly priced ethanol may be possible in a few years, just as flex-fuel vehicles arrive en masse. While the combination of these cars and cheaper ethanol would only begin to reduce our perilous reliance on foreign oil, it would offer fringe benefits, such as lowering U.S. emissions that contribute to global warming. Another plus: Our corn supply is pretty secure—Saddam Hussein isn't likely to invade Iowa anytime soon.

"'Biomass'—corn stalks, rice straw, even household food scraps—could be the source of an alcohol-based 'fuel of tomorrow.'"

The New Alchemy: Turning Garbage into Fuel

John J. Fialka

In the following viewpoint John J. Fialka explains that fibers from sugar cane and household garbage can be used to produce fuel for automobiles. The nation's growing mountains of garbage have encouraged commercial ventures to develop ways to use the refuse to make gasoline alternatives. In addition, strict environmental restrictions on the burning of fossil fuels have made renewable and clean-burning energy from biomass an attractive energy alternative. John J. Fialka is a staff reporter for the *Wall Street Journal*.

As you read, consider the following questions:
1. What is bagasse, according to Fialka?
2. According to Fialka, what percentage of the nation's energy is currently derived from biomass?
3. What have the enzymes from jungle rot been used for, according to the author?

D oes it pay to turn garbage into a substitute for gasoline? The U.S. government has spent almost 50 years and hundreds of millions of dollars trying to prove that it does. The research has produced many rosy policy statements about how the various sorts of organic refuse known as "biomass"—corn stalks, rice straw, even household food scraps—could be the source of an alcohol-based "fuel of tomorrow." But tomorrow never quite seemed to get here.

Now, at least four commercial ventures are gearing up to make gasoline substitutes out of stuff that people throw away or pay others to get rid of. The appearance of economically viable companies outside the government research umbrella is a radical change, says James Hettenhaus, a Charlotte, N.C., consultant who works with the Department of Energy and private companies to promote the technology.

Environmental laws, the threat of future restrictions and the nation's growing mountains of garbage are creating the new marketplace for biomass. For example, research spurred by climate change has begun to convince farmers that they can make more money and make their soil more fertile with new "no till" planting methods, which make plowing unnecessary. That will leave the upper Midwest strewn with some 250 million tons of corn stalks and other agricultural wastes, much of which farmers used to plow under.

"It could be a new cash crop," says Mr. Hettenhaus, who has begun to organize farm groups to collect the stuff, bale it and sell it.

Turning Trash into Ethanol

One buyer he is working with is Iogen Corp., of Ottawa. Iogen is building a refinery there to turn corn stalks into ethanol, which then will be blended with gasoline and sold in the area by the Calgary oil-refiner Petro-Canada. According to Patrick Foody, Iogen's 69-year-old founder, the plant will begin producing fuel this spring.

In Louisiana, bagasse—the waste fibers left after the juice is squeezed out of sugar cane—has always posed a disposal problem. But a Dedham, Mass., company, BC International Corp., has a solution. It has bought a Louisiana refinery that was making ethanol out of corn and is modifying it to break

197

down sugar-cane wastes with a special process that uses genetically altered bacteria and enzymes.

"Bagasse is a no-cost feedstock," says Stephen Gatto, the company's president and chief executive. The company will extract and ferment the sugar from bagasse into ethanol and burn another chemical derived from the process—lignin—as a fuel to run the refinery. Initially the company will sell its ethanol to companies that will process it into mouthwash, hairspray and solvents. Blended gasoline, he says, is a likely future product because there is a federal excise-tax exemption for manufacturers of ethanol-blended fuel in effect until 2007.

"Beyond that there needs to be a change in economics to support this industry," Mr. Gatto says. Biotechnology is producing so many breakthroughs in fuel-production that economic supports one day won't be necessary, he predicts. "Ultimately," he says, "we need to get there."

In California, rice farmers traditionally would get rid of rice straw by burning it in the fields, but new air pollution regulations ban burning. A Mission Viejo company, Arkenol Inc., is building a "biorefinery" near Sacramento to process rice straw to yield ethanol that will be blended with gasoline. Thanks to clean-air regulations, says Russ Miller, the company's chief operating officer, raw material abounds.

In upstate New York, the raw material will be household garbage. Cities such as Middletown are discovering that it is increasingly expensive and difficult to find land for landfills.

Sewage to Fuel the World?

Enter the Masada Resource Group of Birmingham, Ala. It has a contract with Middletown to produce ethanol out of sewage sludge and organic waste gleaned from the city's garbage.

"This will definitely be a commercial facility," predicts David Webster, the project's manager. Seventy percent of Masada's revenue will come from "tipping fees," or the price people pay the company to take their garbage. After plastics, glass and metals are removed, what remains will be shredded, dried and then bathed in acid to remove the sugars.

"The sugar stream continues on into our ethanol production side," explains Mr. Webster. "It is concentrated and then you just go into fermentation, add some yeast and all of a sudden you get alcohol. People have done this since biblical times." While it sees the process as being primarily a "refuse elimination system," Masada also will market the ethanol and carbon dioxide it derives, he says.

At the moment, only about 3% of the nation's energy is derived from biomass, nearly all of it from making ethanol derived from corn or from burning wood. But President Clinton says the nation needs to rely more on biomass in order to cut reliance on oil imports and reduce the threat of global warming.

One of the early pioneers in biomass research was an intrepid Army microbiologist named Elwyn T. Reese. Mr. Reese became intrigued with "jungle rot," a strange fungus that ate the uniforms off U.S. troops in World War II.

Recycling Garbage into Energy

Landfill gas-to-energy projects [which turn the gas that refuse produces as it decomposes into energy] . . . eliminate detrimental air emissions; prevent landfill methane from contributing to global climate changes; prevent methane from migrating off-site and becoming a safety hazard or odor problem; and provide local utilities, industries and consumers with a competitive source of energy. In other words, landfill gas-to-energy facilities provide a unique form of "recycling." Refuse is hauled to the landfill as waste and returned to the consumer in the form of energy.

Tom Kerr, *Corrections Today*, August 1997.

Mr. Reese began his work in the 1950s by isolating the fungus from a rotting pup tent shipped from Bougainville Island in the South Pacific at the end of the war. His bosses at the Army's Research and Development Command, in Natick, Mass., told him the fungus was a threat to the nation's security and to kill it.

But Mr. Reese felt sure there would be commercial uses for it: The strange fungus produced a whole family of enzymes that broke down plants' cellulose structure into sugars more quickly than anything he'd ever seen.

His research gained new impetus during the energy crisis

of the 1970s, as the federal government and private companies scrambled to develop a substitute for gasoline. The Department of Energy (DOE) took over Mr. Reese's research in 1980, working with Gulf Oil at a small refinery that was experimenting with ethanol derived from waste.

Jungle Rot to Jeans

There were problems. Enzymes from jungle rot refused to work when foreign bacteria were around. And they were picky eaters, working efficiently only when agricultural wastes were ground into tiny pieces. This is where Iogen's Mr. Foody came in. He had been using a steam oven to explode wheat straw into smithereens to make it useable as cattle feed. His oven, DOE found, made the wastes more appetizing for the fungus.

Gulf, and later the DOE, lost interest in the research, but Mr. Foody didn't. He discovered that makers of "stonewashed" jeans were interested in enzymes. Last year, manufacturers of over 100 million pairs of jeans used Iogen enzymes to create the faded-denim look, explains Brian Foody, Mr. Foody's son and the company's president. "We had a whole pile of money tied up in this [enzyme-making] equipment," he explains. "We had to find a way to make it a going business."

By the early 1990s, global warming had brought U.S. government researchers back to fungus, as scientists touted it as an environmentally friendly fuel. Amoco invested $7 million dollars in Iogen, but then, in a 1994 reorganization, decided that fungus didn't fit with the corporate culture. That left the Foodys with a new, Amoco-built lab.

Patrick Foody, the Canadian engineer who has built Iogen from the cast-off machinery and fungus research of the U.S. energy research effort, remains optimistic. "Two years from now," he says, "we'll be going head-to-head with gasoline at the pumps."

Periodical Bibliography

The following articles have been selected to supplement the diverse views presented in this chapter. Addresses are provided for periodicals not indexed in the *Readers' Guide to Periodical Literature*, the *Alternative Press Index*, the *Social Sciences Index*, or the *Index to Legal Periodicals and Books*.

Better Homes and Gardens	"Hydrogen: Fuel of the Future?" December 1999.
Choice	"Car of the Future?" October 2000.
Seth Dunn	"Toward a New Energy Ethic," *Enough!*, Spring 2000.
Gregg Easterbrook	"The Energy Trap," *Los Angeles Times*, March 25, 2001.
Economist	"At Last, the Fuel Cell," October 25, 1997.
Catherine Greenman	"Bike Power: Tired Legs Get a Leg Up," *New York Times*, April 19, 2001.
Frank Lingo	"Time to Switch Fuels," *Progressive Populist*, May 1, 2000.
Timothy Maier	"Make Way for 'Green' Machines," *Insight*, May 18, 1998.
Jason Mark	"Who's in the Driver's Seat?" *Dollars and Sense*, July/August 1998.
Mother Earth	"Four Wheels, No Emissions," December 1999.
Motor Trend	"Hybrid and Electric Vehicles," August 1999.
Randal O'Toole	"The Coming War on the Automobile," *Liberty*, March 1998.
Eric Peters	"Weighted Mission of CAFE's Parity Posse," *Washington Times*, August 27, 1996.
Gary Polakovic	"Firms Told to Resume Making Electric Cars," *Los Angeles Times*, September 9, 2000.
Cherie Winner	"Bike It Instead," *Current Health*, October 1995.

For Further Discussion

Chapter 1

1. Christopher Flavin argues that in order to decrease the consumption of fossil fuels, governments need to offer incentives to companies that pursue alternative energy sources. On the other hand, Thomas Sowell contends that governmental interference in the energy market will make energy production of all kinds too expensive and create artificial shortages. In your opinion, what should government's role be in fostering sound energy and environmental policies? Please explain your answer.

2. C.J. Campbell claims that cheap and accessible oil reserves will continue to decline, creating grave economic and political consequences as nations around the world experience energy scarcity and rising prices. Conversely, Sarah A. Emerson contends that technological innovation will continue to enable oil companies to discover more reserves and extract more oil from them. Do you think it is prudent to depend upon future advances in technology to solve today's most serious environmental and energy-related problems? Please explain.

3. David Cromwell maintains that the oil industry is so powerful that it dominates government energy policy, benefiting oil companies at the expense of the environment. However, Jonathan H. Adler contends that the United States government—at the urging of environmentalists—is considering an energy policy that will require a decrease in fossil fuel consumption. In the contest between environmentalists and the oil industry for control of U.S. energy policy, which side do you think is promoting the most beneficial agenda for the future? Please explain your answer.

4. Bob Herbert argues that energy conservation and the development of clean energy sources can help stop the environmental problems associated with the combustion of fossil fuels. On the other hand, Herbert Inhaber claims that energy conservation always fails because the energy saved by some people will inevitably be consumed by others. Supposing Inhaber is correct, what reasons might there be for you to still pursue energy efficiency? In formulating your answer, consider to what extent the actions of others should determine your behavior.

Chapter 2

1. Patrice Hill argues that opposition to nuclear power has decreased as nuclear power plants continue to run more safely.

Conversely, Christopher Flavin and Nicholas Lenssen contend that public opinion is still against nuclear power, in large part because of past nuclear accidents. In your opinion, does the general public hold an accurate view of the dangers and benefits of nuclear power? When constructing your answer, consider from what sources people get information about nuclear power and to what extent they evaluate what they read.

2. John Ritch uses one high profile example of a nuclear accident—the meltdown at Chernobyl—to support his argument that such accidents are rare and do not accurately represent the risks associated with nuclear power. In contrast, Chris Busby uses several specific examples of radiation poisoning to support his argument that nuclear power poses a serious health risk. In your opinion, which author's use of examples is more convincing? Explain your answer.

3. Mary L. Walker contends that nuclear power is cheaper to produce than energy generated from fossil fuels, but she makes no mention of nuclear power's hidden costs, such as government subsidies, which Laura Maggi writes about in her article. On the other hand, when Maggi argues that nuclear power is expensive, she does not mention any of the positive financial aspects of nuclear power that Walker writes about, such as savings generated by nuclear power's protection of the environment. In your opinion, which author's argument is more seriously undermined by omissions? Point to specific omissions in each article when explaining your answer.

Chapter 3

1. Peter Harper argues that wind power is an ethical energy source because the costs of producing it are borne immediately by its users. Do you agree with Harper that it is immoral to use energy from fossil fuels—such as gasoline for your car—because you do not pay the environmental and human-health costs associated with its production and consumption? Please explain your answer.

2. Barbara Wolcott reports that power from the sun is clean and abundant, but she cautions that solar energy in the past has been expensive—about eighteen thousand dollars for a solar power installation for an individual home. Do you believe that government should grant subsidies—at taxpayers' expense—to individual homeowners for the installation of photovoltaic panels for their homes? Why or why not?

3. William McCall reports that fuel cells are expensive, but he explains that new technology may eventually make them afford-

able for the average consumer. Would you pay more for a fuel-cell powered automobile, washer, or other product that is environmentally-friendly? Why or why not?

4. Richard Monastersky describes how methane hydrates found beneath the ocean floor and in permafrost regions may be used as a future energy source. However, the energy source derived from methane hydrates is natural gas—a fossil fuel—which many environmentalists claim emits harmful greenhouse gases when burned. In your opinion, should scientists continue to research methane hydrates as an energy source? Please explain your answer.

5. Marilyn L. Nemzer, Anna K. Carter, and Kenneth P. Nemzer claim that geothermal energy is a reliable, clean, and abundant energy source. In light of the benefits of geothermal energy, why do you think the United States is still relying so heavily on fossil fuels, which many experts claim pollute the environment when burned and exist only in finite quantities? Please explain your answer.

6. France Bequette describes a tidal power plant in France that uses the energy from tides within an estuary to generate electricity. However, such power plants can disrupt sensitive coastal ecosystems by artificially changing water levels. In your opinion, should tidal power plants be built in estuaries in the United States, regardless of their environmental impact? In formulating your answer, weigh the needs of people for more energy sources against the importance of protecting the environment.

Chapter 4

1. Danylo Hawaleshka contends that alternative vehicles can help solve some of the problems caused by gasoline-powered cars, such as the emission of pollutants that harm the environment. However, Jane Holtz Kay argues that alternative vehicles will not help the environment because they still require sprawl-producing highways and their manufacture generates pollutants. In your opinion, how likely is it that Americans would quit depending on personal vehicles for transportation in an effort to help the environment? When constructing your answer, consider to what extent people should put environmental concerns over their own convenience and pleasure.

2. Tom Barry claims that although electric cars have a limited traveling range, advances in technology will double that range to about 120 miles, making such cars more practical. In contrast, Thomas Sowell asserts that electric cars are impractical for most

transportation needs, making them unattractive to consumers. Taking into account the limitations of electric cars, how might consumers be persuaded to purchase them? In other words, if you wanted to convince one of your friends to buy an electric car, what argument would you use?

3. Donald W. Nauss reports that Chrysler is designing a fuel cell car that converts gasoline to hydrogen. However, Nauss explains that many environmentalists worry that the new car will only delay the use of zero-emission fuel cell cars, which will exacerbate environmental problems caused by auto emissions. In your opinion, do interim solutions like Chrysler's hybrid car help or hurt efforts to protect the environment? Please explain your answer.

4. Steve Spaulding contends that natural gas vehicles are superior to gasoline-powered cars because natural gas—unlike oil—is abundant in the United States and when burned, emits much fewer pollutants. However, natural gas is still a finite resource and cars that run on it are not emission-free. Do you think manufacturers should be building natural gas vehicles or should they try to design vehicles that will one day run on clean and renewable energy sources? Please explain your answer.

5. David Stipp argues that ethanol—a clean, renewable energy source—can reduce U.S. reliance on foreign oil and help reduce auto emissions. However, he notes that very few gas stations sell ethanol because the oil companies own the pumps and have no incentive to sell anything but their own products. To what extent do you think choices in vehicles and fuels are dictated by large oil and car companies? Do you believe that alternatives to gas-powered vehicles would be more common than they are now if these large corporations were not so powerful? Explain.

Organizations to Contact

The editors have compiled the following list of organizations concerned with the issues debated in this book. The descriptions are derived from materials provided by the organizations. All have publications or information available for interested readers. The list was compiled on the date of publication of the present volume; names, addresses, phone and fax numbers, and e-mail and Internet addresses may change. Be aware that many organizations take several weeks or longer to respond to inquiries, so allow as much time as possible.

American Petroleum Institute (API)
1220 L St. NW, Washington, DC 20005
(202) 682-8000
website: www.api.org

The American Petroleum Institute is a trade association representing America's petroleum industry. Its activities include lobbying, conducting research, and setting technical standards for the petroleum industry. API publishes numerous position papers, reports, and information sheets.

American Solar Energy Society (ASES)
2400 Central Ave., Suite G-1, Boulder, CO 80301
(303) 443-3130 • fax: (303) 443-3212
e-mail: ases@ases.org • website: www.ases.org

ASES promotes solar energy. It disseminates information on solar energy to schools, universities, and the community. In addition to the *ASES Publications Catalog*, the society publishes the bimonthly magazine *Solar Today*.

American Wind Energy Association (AWEA)
122 C St. NW, Suite 380, Washington, DC 20001
(202) 383-2500 • fax: (202) 383-2505
e-mail: windmail@awea.org • website: www.awea.org

AWEA supports wind energy as an alternative to current energy sources. It provides federal and state legislators with information on wind as an energy source. Publications include the *AWEA Wind Energy Weekly* and the monthly *Windletter*.

Competitive Enterprise Institute (CEI)
1001 Connecticut Ave. NW, Suite 1250, Washington, DC 20036
(202) 331-1010 • fax: (202) 331-0640
e-mail: info@cei.org • website: www.cei.org

CEI advocates removing government environmental regulations to establish a system in which the private sector is responsible for environment and energy policy. Its publications include the monthly newsletter *CEI UpDate* and numerous reprints and briefs.

Conservation and Renewable Energy Inquiry and Referral Service (CAREIRS)

PO Box 8900, Silver Spring, MD 20907

(800) 523-2929

CAREIRS, a project of the U.S. Department of Energy, answers inquiries and provides referrals and information concerning the use of renewable energy technologies and conservation methods. CAREIRS publishes fact sheets, including *Renewable Energy: An Overview*, *Wind Energy*, and *Solar Energy and Your Home: Questions and Answers*.

Council on Alternative Fuels (CAF)

1225 I St. NW, Suite 320, Washington, DC 20005

(202) 898-0711

CAF is comprised of companies interested in the production of synthetic fuels and the research and development of synthetic fuel technology. It publishes information on new alternative fuels in the monthly *Alternate Fuel News*.

Energy Conservation Coalition (ECC)

1525 New Hampshire Ave. NW, Washington, DC 20036

(202) 745-4874

ECC is a group of public interest organizations that promote energy sufficiency. It supports government policies that encourage energy conservation. ECC publishes *Powerline*, a bimonthly periodical covering consumer issues on energy and utilities.

Geothermal Education Office (GEO)

664 Hilary Dr., Tiburon, CA 94920

(415) 435-4574 • fax: (415) 435-7737

e-mail: geo@marin.org • website: http://geothermal.marin.org

GEO's purpose is to promote public understanding about geothermal resources and their importance in providing clean sustainable energy while protecting the environment. GEO produces and distributes educational materials about geothermal energy to schools, energy and environment educators, libraries, industry, and the public.

Geothermal Resources Council (GRC)
PO Box 1350, Davis, CA 95617
(916) 758-2360
e-mail: grclib@geothermal.org • website: www.geothermal.org

GRC encourages the research, exploration, and development of geothermal energy, and supports legislation and regulations that promote the use of such energy. It provides information to the public and publishes the monthly *GRC Bulletin*.

Greenpeace
702 H St. NW, Washington, DC 20001
(800) 326-0959
website: www.greenpeace.org

Greenpeace opposes nuclear energy and supports ocean and wildlife preservation. The organization sponsors public protests against nuclear energy and other activities it believes harm the environment. It publishes the bimonthly magazine *Greenpeace* and many books and reports, including *Global Warming: The Greenpeace Report* and *Saving Energy Is Saving Lives*.

The Heritage Foundation
214 Massachusetts Ave. NE, Washington, DC 20002
(202) 546-4400 • fax: (202) 546-8328
website: www.heritage.org

The Heritage Foundation is a public policy think tank that advocates that the United States increase domestic oil production. Its publications include the quarterly magazine *Policy Review*, brief *Executive Memorandum* editorials, and the longer *Backgrounder* studies.

International Association for Hydrogen Energy (IAHE)
PO Box 248266, Coral Gables, FL 33124
(305) 284-4666
website: www.iahe.org

The IAHE is a group of scientists and engineers professionally involved with the production and use of hydrogen. It sponsors international forums to further its goal of creating an energy system based on hydrogen. The IAHE publishes the monthly *International Journal for Hydrogen Energy*.

National Coal Association (NCA)
1130 17th St. NW, Washington, DC 20036
(202) 463-2653 • fax: (202) 833-1965

NCA is a national trade association that represents the coal industry. The association is primarily a lobbying organization that advocates the use of coal to meet America's energy needs. It publishes the weekly *Coal News* newsletter, the bimonthly magazine *Coal Voice*, and periodic forecasts, fact sheets, and monographs.

Natural Resources Defense Council (NRDC)
40 W. 20th St., New York, NY 10011
(212) 727-2700 • fax: (212) 727-1773
e-mail: nrdcinfo@nrdc.org • website: www.nrdc.org

The council is a nonprofit activist group composed of scientists, lawyers, and citizens who work to promote environmentally safe energy sources and protection of the environment. NRDC publishes a quarterly, the *Amicus Journal*, the newsletter *Newsline*, and a bibliography of books concerning environmental issues.

Political Economy Research Center (PERC)
502 S. 19th Ave., Bozeman, MT 59718
(406) 587-9591
e-mail: perc@perc.org • website: www.perc.org

PERC is a research and education foundation that focuses primarily on environmental and natural resource issues. Its approach emphasizes the use of the free market and the importance of private property rights in protecting the environment and finding new energy resources. Publications include *PERC Viewpoint* and *PERC Reports*.

Renewable Fuels Association (RFA)
1 Massachusetts Ave. NW, Suite 820, Washington, DC 20001
(202) 289-3835 • fax: (202) 289-7519
e-mail: info@ethanolrfa.org • website: www.ethanolrfa.org

RFA is comprised of professionals who research, produce, and market renewable fuels, especially alcohol fuels. It also represents the renewable fuels industry before the federal government. RFA publishes the monthly newsletter *Ethanol Report*.

Union of Concerned Scientists (UCS)
26 Church St., Cambridge, MA 02138
(617) 547-5552
website: www.ucsusa.org

The Union of Concerned Scientists is an organization of scientists and other citizens concerned about nuclear energy and the impact of advanced technology on society. The UCS conducts independent research, sponsors and participates in conferences and panels, and testifies at congressional and regulatory hearings. The UCS publishes a quarterly newsletter, *Nucleus*, as well as books, reports, and briefing papers.

United States Council for Energy Awareness (USCEA)
1776 I St. NW, Washington, DC 20006-2495
(202) 293-0770

The United States Council for Energy Awareness is the public relations organization for the commercial nuclear energy industry in the United States. Its activities include public and media relations and public opinion research. USCEA's publications include the monthly newsletter *INFO*, the quarterly *Nuclear Industry* magazine, and position papers, fact sheets, and brochures.

Worldwatch Institute
1776 Massachusetts Ave. NW, Washington, DC 20036
(202) 452-1999 • fax: (202) 296-7365
e-mail: worldwatch@worldwatch.org
website: www.worldwatch.org

The Worldwatch Institute is a nonprofit research organization that analyzes and focuses attention on global problems, including environmental and energy concerns. The institute, which is funded by private foundations and United Nations organizations, publishes the bimonthly *World Watch* magazine and the Worldwatch Paper series, including *Air Pollution, Beyond the Petroleum Age: Designing a Solar Economy, Alternatives to the Automobile: Transport for Livable Cities*, and *Slowing Global Warming: A Worldwide Strategy.*

Bibliography of Books

Jonathan H. Adler   *Environmentalism at the Crossroads*. Rockville, MD: Government Institutes, 1997.

Peter Asmus   *Reaping the Wind: How Mechanical Wizards, Visionaries, and Profiteers Helped Shape Our Energy Future*. Washington, DC: Island Press, 2000.

Stephen E. Atkins   *Historical Encyclopedia of Atomic Energy*. Westport, CT: Greenwood Publishing, 2000.

Paul Ballonoff   *Energy: Ending the Never-Ending Crisis*. Washington DC: Cato Institute, 1997.

Daniel Berman and John O'Connor   *Who Owns the Sun?* New York: Chelsea Green Publishing, 1997.

Godfrey Boyle, ed.   *Renewable Energy: Power for a Sustainable Future*. New York: Oxford University Press, 1996.

Paul Brown   *Energy and Resources: Living for the Future*. Danbury, CT: Franklin Watts, 1998.

Daniel D. Chiras   *The Natural House: A Complete Guide to Healthy, Energy-Efficient, Environmental Homes*. New York: Chelsea Green, 2000.

Gale E. Christianson   *Greenhouse: The 200-Year Story of Global Warming*. New York: Penguin USA, 2000.

Judith Condon   *Chernobyl and Other Nuclear Accidents*. Austin, TX: Raintree/Steck Vaughn, 1998.

Michael J. Daley   *Nuclear Power: Promise or Peril?* Minneapolis, MN: Lerner Publications, 1997.

David Lewis Feldman   *The Energy Crisis: Unresolved Issues and Enduring Legacies*. Baltimore, MD: Johns Hopkins University Press, 1996.

John Fenton and Ron Hodkinson   *Lightweight Electric/Hybrid Vehicle Design*. Hillsboro, OR: Butteworth-Heinemann, 2001.

Christopher Flavin and Nicholas Lenssen   *Power Surge: Guide to the Coming Energy Revolution*. New York: W.W. Norton, 1994.

Ross Gelbspan   *The Heat Is On*. New York: Perseus Press, 1998.

Paul Gipe   *Wind Energy Basics*. New York: Chelsea Green Publishing, 1999.

John G. Ingersoll   *Natural Gas Vehicles*. Lilburn, GA: Fairmont Press, 1995.

Herbert Inhaber   *Why Energy Conservation Fails*. Westport, CT: Greenwood Publishing Group, 1997.

James Larminie and Andrew Dick   *Fuel Cells Explained*. Etobicoke, Ontario: John Wiley and Sons, 2000.

James Manwell, Jon McGowan and Anthony Rogers — *Wind Energy Explained*. Etobicoke, Ontario: John Wiley and Sons, 2001.

Tomas Markvart, ed. — *Solar Electricity*. Etobicoke, Ontario: John Wiley and Sons, 2000.

Robert C. Morris — *The Enviromental Case for Nuclear Power: Economic, Medical, and Political*. St. Paul, MN: Paragon House, 2000.

Jim Motavalli — *Forward Drive: The Race to Build the Car of the Future*. San Francisco: Sierra Club Books, 2000.

Mukund R. Patel — *Wind and Solar Power Systems*. Boca Raton, FL: CRC Press, 1999.

Frederico Pena — *Turning Off the Heat: Why America Must Double Energy Efficiency to Save Money and Reduce Global Warming*. Amherst, NY: Prometheus Books, 1998.

Robert W. Righter — *Wind Energy in America: A History*. Norman: University of Oklahoma Press, 1996.

Robert A. Ristinen and Jack I. Kraushaar — *Energy and the Environment*. Etobicoke, Ontario: John Wiley and Sons, 1998.

Max Roehr, ed. — *The Biotechnology of Ethanol: Classical and Future Applications*. Etobicoke, Ontario: John Wiley and Sons, 2001.

David Ross — *Power from the Waves*. New York: Oxford University Press, 1994.

Gunter Simader and Karl Kadesch — *Fuel Cells and Their Applications*. Etobicoke, Ontario: John Wiley and Sons, 1996.

Madanjeet Singh — *Timeless Energy of the Sun: For Life and Peace with Nature*. San Francisco: Sierra Club Books, 1998.

Robert Snedden — *Energy Alternatives*. Westport, CT: Heinemann Library, 2001.

Ernest Henry Wakefield — *History of the Electric Automobile: Hybrid Electric Cars*. New York: Society of Automotive Engineers, 1998.

Thomas Raymond Wellock — *Critical Masses: Opposition to Nuclear Power in California*. Madison: University of Wisconsin Press, 1998.

James C. Williams — *Energy and the Making of Modern California*. Akron, OH: University of Akron Press, 2001.

Donald R. Wulfinghoff — *Energy Efficiency Manual*. Washington, DC: Energy Institute Press, 2000.

Index